INSTRUMENTATION AND MEASUREMENT IN ELECTRICAL ENGINEERING

INSTRUMENTATION AND MEASUREMENT IN ELECTRICAL ENGINEERING

ROMAN MALARIĆ

BrownWalker Press
Boca Raton

Instrumentation and Measurement in Electrical Engineering

Copyright © 2011 Roman Malarić
All rights reserved.
No part of this book may be reproduced or transmitted in any form or by any means, electronic or mechanical, including photocopying, recording, or by any information storage and retrieval system, without written permission from the publisher.

BrownWalker Press
Boca Raton, Florida
USA • 2011

ISBN-10: 1-61233-500-4 *(paper)*
ISBN-13: 978-1-61233-500-1 *(paper)*

ISBN-10: 1-61233-501-2 *(ebook)*
ISBN-13: 978-1-61233-501-8 *(ebook)*

www.brownwalker.com

Some of the electrical symbols used in this book

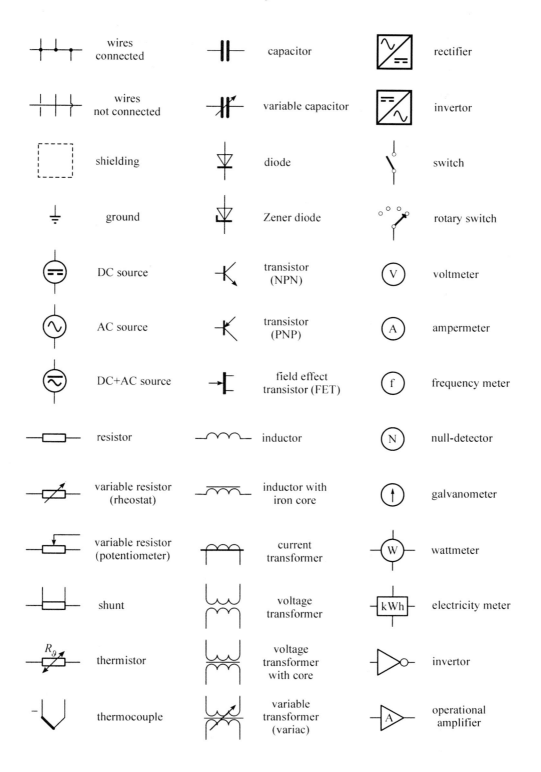

PREFACE

The inclusion of an electrical measurement course in the undergraduate curriculum of electrical engineering is important in forming the technical and scientific knowledge of future electrical engineers. This book explains the basic measurement techniques, instruments, and methods used in everyday practice. It covers in detail both analogue and digital instruments, measurements errors and uncertainty, instrument transformers, bridges, amplifiers, oscilloscopes, data acquisition, sensors, instrument controls and measurement systems. The reader will learn how to apply the most appropriate measurement method and instrument for a particular application, and how to assemble the measurement system from physical quantity to the digital data in a computer. The book is primarily intended to cover all necessary topics of instrumentation and measurement for students of electrical engineering, but can also serve as a reference for engineers and practitioners to expand or refresh their knowledge in this field.

ACKNOWLEDGMENTS

I would like to thank Ivica Kunšt, dipl. ing for his suggestions and for designing most of the figures in this book. I also wish to thank my colleagues at the Faculty, as well as my colleagues from the TEMUS-158599 project "Creation of the Third Cycle of Studies – Doctoral Studies in Metrology" for their support. Special thanks go to my mother Marija, father Vladimir, and my brother Krešimir for their encouragement and assistance. And finally, thanks to my wife Božica and to my kids for their patience and support.

INTRODUCTION

Measurement followed man from the very beginning of its development. Measuring methods and measuring instruments were developed in parallel with the development of electrical engineering. However, some physical laws were derived based on measurement results, such as the Biot-Savart law, when in 1820, the French scientists Jean-Baptiste Biot and Félix Savart established the relationship between an electric current and the magnetic field it produces. Although science and metrology (the science of measurements) are developing quickly, one should always remember that a measurement principle established more than 150 years ago can still be applicable today. There will be many such principles explained in this book. As the instrumentation becomes more advanced, results will only become more precise.

The basic purpose of metrology is best described by the famous Italian scientist Galileo Galilei: "Measure everything that can be measured and try to make measurable what is not yet measurable." The term *metrology* is derived from the Greek words *metron*–to measure–and *logos*– science. The process of measurement involves comparison of the measured quantity with the specific unit; it is therefore necessary to know the unit of measurement with the highest possible accuracy. The first modern metrology institute was established in 1887 in Germany. This institute was partly responsible for the sudden rise of strength of German industry in the world. Very soon thereafter, other industrial countries established metrology institutes in order to maintain their places at the top of world industry. With the progress of science foundation around the world, metrology slowly relied more and more on natural phenomena and not on prototypes, as in the past. Today, the only unit of measurement embodied in prototype–kilogram–is stored in Sèvres near Paris, but in recent years, the metrology world appears to desire to define this unit by natural phenomena just like all the others.

Today, metrology is based on natural laws and is unique in how the units can be realized anywhere in the world, if only one has the necessary knowledge and equipment. The foundation of today's metrology is the International System of Units (SI), adopted in 1960. This system consists of seven base units and a large number of derived units, 23 of which have their own special names and signs. Only electrical current is included in the seven base units from electrical engineering. All others, such as units for electrical resistance and voltage, are derived units.

Chapter 1 gives an overview of the modern SI system of units, and explains the definition of SI base units, its realization, and its standards.
Chapter 2 describes measurement errors, calculation of measurement uncertainty, and instrumentation limits of errors.
Chapter 3 describes the different measuring elements such as resistors, inductors, and capacitors, as well as voltage standards.
Chapter 4 describes the analogue measuring instruments, and how to use different types to measure various AC and DC voltages and currents.
Chapter 5 is about the compensation measurement methods, such as bridges and compensators. AC and DC calibrators are also described.

Chapter 6 gives an overview of instrument transformers, their uses, and testing methods for determination of phase and current/voltage errors.

Chapter 7 describes the use of operation amplifiers in measurement technology, and how to use them to build electronic instruments and other devices using the op-amps.

Chapter 8 gives an overview of cathode ray tube and digital storage oscilloscopes. How to use oscilloscopes to measure different electrical quantities is also described.

Chapter 9 describes the construction and use of digital multimeters, and provides an overview of different analogue to digital converters used in various digital instruments.

Chapter 10 describes measurement methods to measure different electrical quantities, such as voltage, current, resistance capacitance, and many others.

Chapter 11 describes the historical development of instrument control with the use of the personal computer, as well as different connectivity interface buses and software support for both stand-alone and modular instrumentation.

Chapter 12 gives an overview of the measurement system, and describes the most popular sensors, signal conditioning, and data acquisition hardware.

TABLE OF CONTENTS

Preface VII
Acknowledgments IX
Introduction XI
Acronyms and Abbreviations XIX

1. INTERNATIONAL SYSTEM OF UNITS - SI 1
   1.1 DEFINITIONS OF THE SI BASE UNITS 2
   1.2 REALIZATION OF UNITS 5
   1.3 PHYSICAL STANDARDS OF UNITS 6
   1.4 TRACEABILITY 6
   1.5 UNITS OUTSIDE THE SI 7
   1.6 SI PREFIXES 8
   1.7 BINARY UNITS 8

2. MEASUREMENT ERRORS 11
   2.1 GRAVE ERRORS 12
   2.2 SYSTEMATIC ERRORS 12
   2.3 RANDOM ERRORS 12
   2.4 CONFIDENCE LIMITS 16
   2.5 MEASUREMENT UNCERTAINTY 18
   2.6 LIMITS OF ERROR (SPECIFICATIONS) 19
      2.6.1 Indirectly Measured Quantities 20

3. MEASURING ELEMENTS 23
   3.1 RESISTORS 23
      3.1.2 Resistance Decades and Slide Resistors 24
      3.1.3 Resistance Standards 25
      3.1.4 Oil Bath 27
      3.1.5 Group Standards 29
      3.1.6 Hamon Transfer Resistor 33
      3.1.7 Quantum Hall Resistance Standard 34
   3.2 CAPACITORS 35
      3.2.1 Equivalent Circuit of Capacitors 35
      3.2.2 Capacitor Standards 36
   3.3 INDUCTORS 37
      3.3.1 Equivalent Circuit of Inductors 38
      3.3.2 Inductance Standards 38
   3.4 LABORATORY VOLTAGE SOURCES 39
   3.5 VOLTAGE STANDARDS 39
      3.5.1 Josephson Array Voltage Standard 39
      3.5.2 Weston Voltage Standard 40
      3.5.3 Electronic Voltage Standards 42
   3.6 ADJUSTING THE CURRENT 43
      3.6.1 Adjusting the Current with Potentiometer 43
      3.6.2 Adjusting the Current with Resistor 44

4. ANALOGUE MEASURING INSTRUMENTS 47
4.1 BASIC CHARACTERISTICS OF ANALOGUE INSTRUMENTS 47
4.1.1 Torque and Counter-Torque 47
4.1.2 The Scale and Pointing Device 48
4.1.3 Uncertainty in Reading Analogue Instruments 49
4.1.4 Sensitivity and Analogue Instrument Constant 50
4.1.5 Standards and Regulations for the Use of Analogue Electrical Measuring Instruments 50
4.2. INSTRUMENT WITH MOVING COIL AND PERMANENT MAGNET (IMCPM) 52
4.2.1 Extending the Measurement Range 53
4.2.2 Measurement of Alternating Current and Voltage Using IMCPM 54
4.2.3 Universal Measuring Instruments 57
4.3 INSTRUMENT WITH MOVING IRON 57
4.4 ELECTRODYNAMIC INSTRUMENT 58
4.5 ELECTRICITY METERING 59
4.5.1 Induction Meter 59
4.5.2 Electricity Meter Testing 62

5. BRIDGES AND CALIBRATORS 65
5.1 DC BRIDGES 65
5.1.1 Wheatstone Bridge 65
5.1.2 Sensitivity of Wheatstone bridge 67
5.1.3 Partially Balanced Wheatstone Bridge 68
5.1.4 Thompson Bridge 68
5.2 AC WHEATSTONE BRIDGE 69
5.3 DC COMPENSATION METHODS 70
5.4 CALIBRATORS 72
5.4.1 DC Calibrator 72
5.4.2 AC Calibrator 74

6. INSTRUMENT TRANSFORMERS 77
6.1 CONNECTING INSTRUMENT TRANSFORMERS 77
6.2 IDEAL AND REAL TRANSFORMERS 77
6.3 VOLTAGE INSTRUMENT TRANSFORMER 79
6.4 CAPACITIVE MEASURING TRANSFORMERS 81
6.5 CURRENT INSTRUMENT TRANSFORMER 82
6.6 CURRENT INSTRUMENT TRANSFORMER ACCURACY TESTING 85
6.6.1 Schering and Alberti Method 85
6.6.2 Hohle Method 86
6.6 WINDING CONFIGURATIONS 87

7. AMPLIFIERS IN MEASUREMENT TECHNOLOGY 89
7.1. MEASURING AMPLIFIERS 89
7.2. OPERATIONAL AMPLIFIERS 92
7.3 OPERATIONAL AMPLIFIER APPLICATIONS 93
7.3.1 Inverting Amplifier 93
7.3.2 Summing Amplifier 94
7.3.3 Non-Inverting Amplifier 94
7.3.4 Integrating Amplifier 94
7.3.5 Differentiator Amplifier 95
7.3.6 Logarithmic Amplifier 95
7.3.7 Voltage Follower 96
7.3.8 Difference Amplifier 96

    7.3.9 Instrumentation Amplifier 97
    7.3.10 Active Guard 97
    7.3.11 Current to Voltage Converter (Transimpedance Amplifier) 98
    7.3.12 Voltage to Current Converter (Transconductance Amplifier) 98
  7.4. MEASURING INSTRUMENTS USING OPERATIONAL AMPLIFIERS 99
    7.4.1. DC Electronic Voltmeters 99
    7.4.2 AC Electronic Voltmeters 100
    7.4.3 AC Voltmeters with Response to the Effective Value 103

8. OSCILLOSCOPES 107
  8.1. CATHODE RAY TUBE 108
  8.2. SYSTEM FOR VERTICAL DEFLECTION 109
  8.3. SYSTEM FOR HORIZONTAL DEFLECTION 112
  8.4. DESCRIPTION FRONT OSCILLOSCOPE PANEL (TEKTRONIX 2205) 114
  8.5. MEASUREMENT USING OSCILLOSCOPES 116
  8.6 DIGITAL STORAGE OSCILLOSCOPES (DSO) 116
    8.6.1 Sampling Methods 117

9. DIGITAL INSTRUMENTS 121
  9.1 ANALOGUE TO DIGITAL CONVERTERS 124
    9.1.1 A/D Converter of Voltage to Time 124
    9.1.2 Dual Slope (Integrating) A/D Converter 125
    9.1.3 Successive Approximation A/D Converter 126
    9.1.4. Parallel A/D Converter 127
  9.2 AC MEASUREMENT IN DIGITAL MULTIMETERS 128
  9.3 MAIN CHARACTERISTICS OF DIGITAL INSTRUMENTS 130
  9.4 ELECTRONIC WATTMETER 130
  9.5 ELECTRONIC ELECTRICITY METERS 132

10. MEASUREMENT OF ELECTRICAL QUANTITIES 135
  10.1 VOLTAGE AND CURRENT MEASUREMENTS 135
    10.1.1 Measurement of Small Currents and Voltages 135
    10.1.2 Measurement of Large Currents 136
    10.1.3 Measurement of High Voltages 137
  10.2 POWER MEASUREMENT 140
    10.2.1 Measurement of DC Power 141
    10.2.2 Measurement of Power Using Watt-meters 142
    10.2.3 Connecting the Wattmeter 143
    10.2.4 Three Voltmeter Method 144
    10.2.5 Three Ammeters Method 145
    10.2.6 Measurement of Active Power in Three-Phase Systems 146
    10.2.7 Aron Connection 148
  10.3 RESISTANCE MEASUREMENT 149
    10.3.1 Voltmeter-Ammeter Method 150
    10.3.2 Compensation and Digital Voltmeter Methods 151
    10.3.3 Measuring Resistance Using the Ohm-Meter 152
    10.3.4 Digital Ohm-Meter 153
    10.3.5 Measurement of Insulation Resistance 154
    10.3.6 Measurement of High-Ohm Resistance 154
    10.3.7 Measurement of Earth Resistance 156
    10.3.8 Measurement of Soil Resistivity 157
  10.4 MEASUREMENT OF IMPEDANCE 158

10.5 MEASUREMENT OF INDUCTANCE 158
   10.5.1 Bridge with Variable Inductance 159
   10.5.2 Maxwell Bridge 160
10.6 CAPACITANCE MEASUREMENT 161
   10.6.1 Wien Bridge 161
   10.6.2 Schering Bridge 162
   10.6.3 Transformer Bridges 165
10.7 MEASURING IMPEDANCE BY SELF-ADJUSTING BRIDGE 166
10.8 TIME, FREQUENCY AND PERIOD MEASUREMENTS 168

11. INSTRUMENTATION AND COMPUTERS 171
11.1 HISTORY OF INSTRUMENTATION AND COMPUTERS, INTERFACES AND BUSES 171
11.2. INTERFACE BUSES FOR STANDALONE INSTRUMENTS 172
   11.2.1 General Purpose Interface Bus (GPIB) 172
   11.2.2 IEEE 488.2 Standard 173
   11.2.3 HS488 175
   11.2.4 RS-232 and RS-485 Serial Connection 175
   11.2.5 Ethernet and VXI-11 Standard 176
   11.2.6 LXI – LAN Extensions for Instrumentation 176
   11.2.7 Universal Serial Bus (USB) and USTMC Class 177
   11.2.8 IEEE 1394 177
   11.2.9 Comparison of Interface Buses for Standalone Instruments 178
11.3. INTERFACE BUSES FOR MODULAR INSTRUMENTS 179
   11.3.1 Peripheral Component Interconnect (PCI) and PCI Express Bus 179
   11.3.2 VXI (VMEbus eXtensions for Instrumentation) Bus 179
   11.3.3 PCI eXtensions for Instrumentation (PXI) 180
   11.3.4 Wireless Connectivity in Measurement Applications 181
11.4 SOFTWARE SUPPORT FOR INSTRUMENT CONTROL 181
   11.4.1 Virtual Instrumentation 182
   11.4.2 Graphical Programming 182
   11.4.3 LabVIEW™ Graphical Programming Language 182
   11.4.4 IEEE 488.2 185
   11.4.5 Standard Commands for Programmable Instruments 186
   11.4.6 Virtual Instrument Software Architecture (VISA) 187
   11.4.7 Instrument Drivers 189

12. MEASUREMENT SYSTEMS 193
12.1 MEASUREMENT SYSTEMS OVERVIEW 193
12.2 SENSORS AND TRANSDUCERS 193
   12.2.1 Thermocouples 194
   12.2.2 Resistance Thermal Devices (RTDs) 195
   12.2.3 Thermistors 197
   12.2.4 Strain Gauges 197
   12.2.5 Linear Voltage Differential Transformer (LVDT) 199
   12.2.6 Potentiometers as Displacement Sensors 200
   12.2.7 Accelerometers 200
   12.2.8 Micro Machined Inertial Sensors (MEMS) 201
12.3 TYPES OF SIGNALS 202
12.4 SIGNAL CONDITIONING 203
   12.4.1 Amplification 203
   12.4.2 Excitation 203
   12.4.3 Linearization 204

 12.4.4 Isolation..204
 12.4.5 Filtering...205
 12.4.6 Comparison of Signal Conditioning Requirements for Different Sensors...........205
12.5 DATA ACQUISITION HARDWARE 206
 12.5.1 DAQ Characteristics..206
 12.5.2 Grounding Issues of DAQ Measurement System...208
 12.5.3 Sources of Noise in the DAQ Measurement System...210
12.6 EMBEDDED AND SYSTEM ON CHIP (SoC) MEASUREMENT SYSTEM 214
 12.6.1 Smart Sensors 214
 12.6.2 Wireless Sensor Networks 215

About the Author..217
Index..219

Acronyms and Abbreviations

A/D	analogue to digital converter
AC	alternate current
ADC	analogue to digital converter
API	Application Programming Interfaces
BIPM	International Bureau of Weights and Measures
CGPM	General Conference on Weights and Measures
CIPM	International Committee for Weights and Measures
CMRR	common-mode rejection ratio
CMV	common-mode voltage
CPU	central processing unit
DAQ	digital acquisition
DC	direct current
DMM	digital multimeter
DSO	digital storage oscilloscope
DVD	cgital video disk
DVM	digital voltmeter
EA	European cooperation for Accreditation
EMF	electromagnetic force
FER	Faculty of Electrical Engineering and Computing in Zagreb
FFT	Fast Fourier Transform
GND	ground
GPIB	General Purpose Interface Bus
HPIB	Hewlett Packard Interface Bus
IEC	International Electrotechnical Commission
IEEE	Institute of Electrical and Electronic Engineers
IICP	Industrial instrumentation control protocol
IMCPM	instrument with moving coil and permanent magnet
ISO	International organization for standardization
JAVS	Josephson Voltage Array standard
LabVIEW	Laboratory virtual instrumentation engineering workbench
LAN	Local Area Network
LCD	Liquid crystal display
LVDT	Linear Voltage Differential Transformer
LXI	LAN Extensions for Instrumentation
MEMS	Micro Machined Inertial Sensors
MOSFET	metal–oxide–semiconductor field-effect transistor
NMI	National Metrology Institute
NRSE	Non-referenced single ended
OP-AMP	operational amplifier
PCI	Peripheral Component Interconnect
PEL	Primary Electromagnetic Laboratory
PTB	Physikalisch Technische Bundesanstalt
PTR	Physikalisch Technische Reichsanstalt
PWMDAC	analogue to digital converter with pulse width modulation
PXI	PCI eXtensions for Instrumentation

RMS	root mean square
RS-232	Recommended Standard 232
RSE	referenced single ended
RTD	Resistance Thermal Devices
SCPI	Standard Commands for Programmable Instruments
SI	Le Système International d'Unités
SoC	System on chip
USB	Universal Serial Bus
USBTMC	USB Test& Measurement Class
VISA	Virtual Instrument Software Architecture
VME	Versa Module Eurocard
VXI	VMEbus eXtensions for Instrumentation
WSN	wireless sensor networks

1. INTERNATIONAL SYSTEM OF UNITS - SI

The International System of Units (SI - Le Système International d'Unités) was established in 1960. It was an important step, after decades of hard work, to overcome the many different units used throughout the world. The need for a unified system of units was evident after the Industrial Revolution in the 18th century. Several important events also contributed to and sped up the process, especially the World Fairs in London (1851) and Paris (1876).

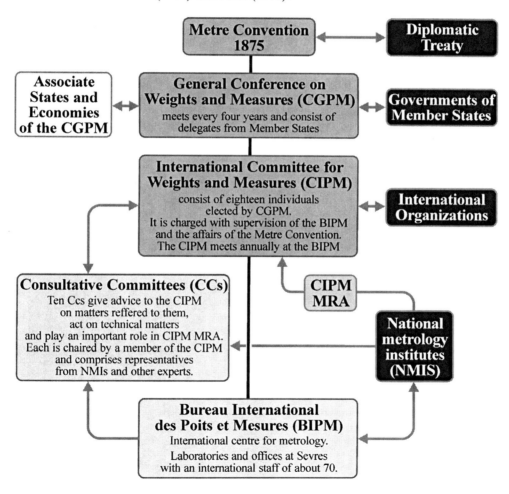

Figure 1.1 Relationships between different metrology organizations in the world

The origin of today's SI system of units goes back to May 20th, 1875, when the representatives of the 17 most technologically-advanced countries of the world signed the Treaty of the Metre (also known as the Metre Convention). At that time three important organizations were created, with the task to take care of the metric standards. These institutions, which are still working today, are:

- International Bureau of Weights and Measures (**BIPM**);
- General Conference on Weights and Measures (**CGPM**);
- International Committee for Weights and Measures (**CIPM**).

The International Committee for Weights and Measures (CIPM) works through ten consultative committees that provide recommendations to the General Conference on Weights and Measures (CGPM), which decides on the resolutions at least once every 6 years. These resolutions are obligatory to the signatories of the Metre Convention (52 countries signed the treaty by the end of 2009). The International Bureau of Weights and Measures (BIPM) is the scientific institute governed by the CIPM. Its role is storing and realizing international primary units, improving measurement methods and units, and comparing different standards for member countries (Figure 1.1).

The SI system of units is a modern metric system, which is used throughout the world today. Even in countries where units that do not belong to SI are used, like in the USA, units are also derived from SI units. It can be stated that the International System of Units (SI) is coherent, unified, and uniform. It is coherent because it is composed of seven base units (meter, kilogram, second, ampere, kelvin, mole, and candela), mutually independent, and units are derived from base units. Coherence means that the basic unit of natural laws is always associated with factor 1 (1x1 = 1, 1/1 = 1; Figure 1.1). It is unified because, except for weight, all units are defined by unchangeable natural constants. It is uniform because the measurements in the dynamics, electrodynamics, and thermodynamics can be compared with each other in terms of conservation of mass and energy (Figure 1.1). The importance of using the SI units is best demonstrated in the unfortunate loss of the **Mars Climate Orbiter** in 1998. The thrusters on the spacecraft, which were intended to control its rate of rotation, were controlled by software that used the unit of **pound force** to make calculations (this is a standard unit for **force** in the United States customary units system). The ratio of SI unit of force **Newton** and the unit of **pound force** is 4.45, and as the spacecraft expected the figures to be in Newtons, the unfortunate mix of units caused the spacecraft to drift into low orbit and be destroyed by atmospheric friction.

1.1 DEFINITIONS OF THE SI BASE UNITS

There is a difference between the unit definition and its realization. The International System of Units (SI) is a set of **definitions**. National Metrology Institutions (NMIs) perform experiments to produce (**realize**) the unit according to the definition, and with some of these experiments the unit can be **stored** in **standards**. The standards are physical objects whose characteristics agree with the **definition** of unit. For example, the unit for **time** is second, and it is defined as the duration of a certain number of periods of the radiation of the atom of cesium-133. Anyone who has the money, knowledge, and equipment can make an atomic clock that produces radiation as defined by the SI unit of second. It is important to emphasize that the atomic clock is not the realization of a second. The realization of a second is the radiation produced by atomic clocks.

There are seven **SI base units** and their **definitions** are:

Unit of length – meter: "The meter is the length of the path traveled by light in vacuum during a time interval of 1/299 792 458 of a second."

***Unit of mass* – kilogram**: "The kilogram is the unit of mass; it is equal to the mass of the international prototype of the kilogram."

***Unit of time* – second**: "The second is the duration of 9 192 631 770 periods of the radiation corresponding to the transition between the two hyperfine levels of the ground state of the cesium 133 atom."

***Unit of electric current* – ampere**: "The ampere is that constant current which, if maintained in two straight parallel conductors of infinite length, of negligible circular cross-section, and placed 1 meter apart in vacuum, would produce between these conductors a force equal to 2 x 10-7 newton per meter of length."

***Unit of thermodynamic temperature* – kelvin**: "The kelvin, unit of thermodynamic temperature, is the fraction 1/273.16 of the thermodynamic temperature of the triple point of water."

***Unit of amount of substance* – mole**: "The mole is the amount of substance of a system which contains as many elementary entities as there are atoms in 0.012 kilogram of carbon 12; its symbol is 'mol.'" When the mole is used, the elementary entities must be specified and may be atoms, molecules, ions, electrons, other particles, or specified groups of such particles.

***Unit of luminous intensity* – candela**: "The candela is the luminous intensity, in a given direction, of a source that emits monochromatic radiation of frequency 540 x 1012 hertz and that has a radiant intensity in that direction of 1/683 watt per steradian."

There are a few dozen **derived** units, some of which are **named** (e.g., **Ohm**, **Volt**), while the majority has **no special name** (e.g. unit for speed is **m/s**).

As already stated, all derived units must be able to be expressed in terms of base units. This will be explained by the derived unit of electrical resistance **Ohm**, named after the German physicist **Georg Simon Ohm** (1787-1854). The unit symbol is **Ω** (capital Greek letter omega -Ω). Ohm is defined as:

$$\Omega = \frac{V}{A}, \qquad [1.1]$$

where **V=W/A**; V is the SI unit of **voltage**, **A (ampere)** is the SI base unit for **current**, and **W (watt)** is the SI unit of **force**. As unit **Watt** is derived from SI units **joule** and **newton**, and those units are derived from base SI units **kilogram**, **meter**, and **second**, the Ohm, expressed by the base units of SI, is:

$$\Omega = m^2 \cdot kg \cdot s^3 \cdot A^2 \qquad [1.2]$$

This is valid for all derived units, which can be seen in Figure 1.2.

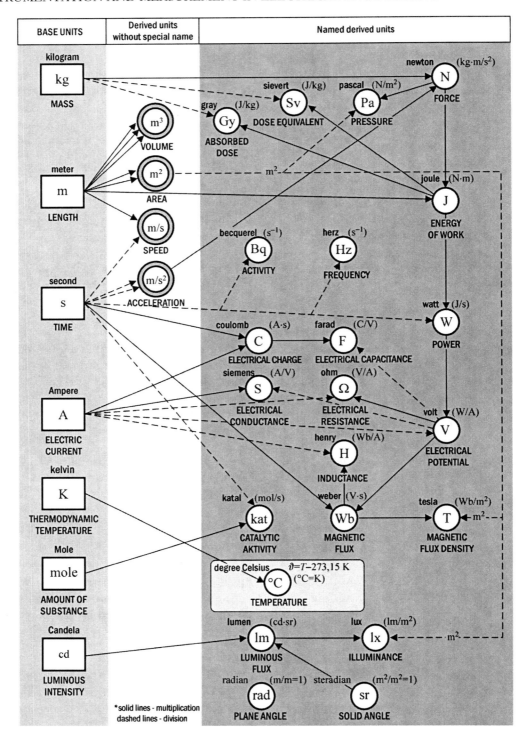

Figure 1.2 SI system – the relationship of base and derived units

1.2 REALIZATION OF UNITS

An example of unit realization is the absolute determination of resistance by the **Thompson-Lampard** calculable capacitance standard. **Calculable standards** are standards whose values can be determined accurately using known dimensions. Such standards should have a simple geometric shape that allows mathematical calculation of standards. This was achieved for the capacitance standard, which was developed by **A.M. Thompson** on theoretical considerations from **D.G. Lampard** in 1956. These considerations show that in the symmetrical arrangement of cylinders A, B, C, and D, according to Figure 1.3, the capacitance between cylinders A and C or B and D is:

$$C_{AC} = C_{BD} = \frac{\varepsilon_0}{\pi}(l_1 - l_2)\ln 2 \qquad [1.3]$$

where ε_0 is the permittivity of vacuum, which can be determined with relative measurement uncertainty of 8×10^9. It is only needed to achieve superb accuracy in measuring the distance between the electrodes, which can be achieved by interference methods, so the total measurement errors of the measurement procedure do not exceed 1×10^7, while the edge effects are avoided by making two measurements at different intervals ($l_1 - l_2$). This realized SI unit of **farad** has become the starting place for the realization of the SI unit of resistance. Realization of the SI unit of resistance from the realized unit of capacitance is shown in Figure 1.4. This entire sequence of ohm realization is achieved with the measurement uncertainty of about 0.1 ppm.

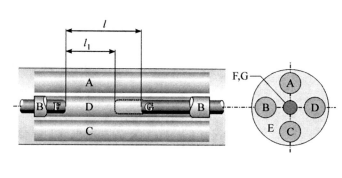

Symmetrical layout, boundary effects are ignored
$$C_{AC} = C_{BD} = \frac{\varepsilon_0}{\pi} l \ln 2$$

Approximately symmetrical layout, boundary effects are ignored
$$\frac{C_{AC} + C_{BD}}{2} = \frac{\varepsilon_0}{\pi} l \ln 2$$

Approximately symmetrical layout, boundary effects are eliminated
$$\frac{\Delta C_{AC} + \Delta C_{BD}}{2} = \frac{\varepsilon_0}{\pi}(l - l_1)\ln 2$$

Figure 1.3 Thompson-Lampard calculable cross capacitor

The unit of volt, for example, is the SI unit defined as the power of one watt divided by the current of one ampere. An example of realization of the unit of volt is the voltage balance, which tries to equal the voltage electrostatic force and mass of the known weight. Results thus obtained are assigned to the standards of cheaper and simpler systems such as **Weston cells** and **Zener voltage standard**, or expensive and complicated devices such as the **Josephson voltage source**, as these devices store and do not realize the SI unit of **volt**, because the way they are used does not include continuous comparison of electrical and mechanical power realized in the system of units SI. Such equipment or devices, primary or secondary, are called **standards**.

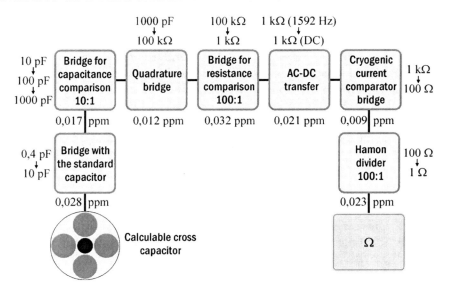

Figure 1.4 Absolute determination of the unit of electrical resistance

1.3 PHYSICAL STANDARDS OF UNITS

Each unit, as stated, has its own definition, realization, and standards. When a certain unit is realized, its value is given to devices that store units - standards. Each national laboratory uses primary or national standards which are used for comparison with other (secondary or working) standards. A standard should ideally have the following characteristics:

- that it is contained in the equipment or device, or that it is easily reproducible in a scientific experiments,
- if it is contained in the device, there must be a way to transmit its values to other standards in different places and from different measurers,
- if it is reproducible in experiments, the same results must still be obtained in all places and under the same conditions,
- there must be a way to transfer it to smaller or larger values with high precision.

National standards have decades of history, so their time stability or drift and uncertainty of the stored units are well-known. There are also so-called **intrinsic** standards for which there is a procedure which, if properly followed, allows getting measurement quantity without additional errors, always within the limits of pre-defined measurement uncertainties. Such standards in electrical measurements are the **Josephson voltage standard**, the **Quantum Hall resistance standard,** and the **Atomic clock time standard**. If the laboratory has an intrinsic standard, there is no need for the services of other laboratories, at least theoretically.

1.4 TRACEABILITY

An important factor in today's international metrology is **traceability**. Traceability is a continuous series of properly conducted and documented comparisons from current measurements through secondary and primary standards to national standards, and eventually to one of the seven SI base units. This system is often referred to as the traceability pyramid (Figure 1.5). Today, traceability is taken for granted, but it was not always this way. The unit of distance *elbow* was variously defined in the Middle Ages. There was, however, the conclusion that it was the length of the index finger to the elbow, but whose elbow? Usually, the ruler's elbow was used. Thus, the measure of distance differed from nation to nation. Today, this problem is avoided with an international agreement on the definition of units. Anyone who wants to have internationally recognized units must accept SI units and ensure their traceability to them.

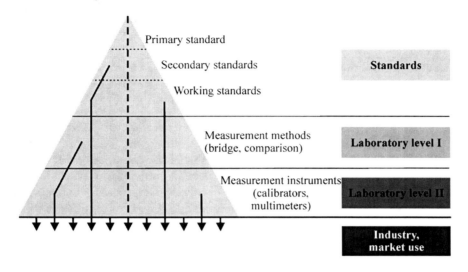

Figure 1.5 Traceability pyramid

1.5 UNITS OUTSIDE THE SI

Some units are not part of the SI system, but because of their importance and for historical reasons, they are still widely used. Some of these important units are listed in Table 1.1.

Table 1.1 Units outside the SI

Name	Symbol	Value in SI units
minute (time)	min	1 min = 60 s
hour (time)	H	1 h = 3600 s
day	D	1 d = 86400 s
degree (angle)	°	$1° = (\pi/180)$ rad
minute (angle)	'	$1' = (1/60)° = (\pi/10800)$ rad
second (angle)	''	$1'' = (1/60)' = (\pi/648000)$ rad
liter	L	$1 L = 1 dm^3 = 10^{-3} m^3$
metric ton	T	$1 t = 10^3$ kg
electron Volt	eV	$1 eV = 1.602\ 18 \times 10^{-19}$ J, approximately
unified atomic mass unit	U	$1 u = 1.660\ 54 \times 10^{-27}$ kg, approximately
astronomical unit	Ua	$1 ua = 1.495\ 98 \times 10^{11}$ m, approximately

There are also some other units currently accepted along with the SI units which are subject to further review by the CIPM (Table 1.2)

Table 1.2 Other non-SI units

Quantity (Name)	Symbol	Value in SI units
pressure (bar)	bar	1 bar = 0.1 MPa = 100 kPa = 10^5 Pa
pressure (millimeter of mercury)	mmHg	1 mmHg ≈ 133.322 Pa
length (ångström)	Å	1 Å = 0.1 nm = 100 pm = 10^{-10} m
distance (nautical mile)	M	1 M = 1852 m
area (barn)	b	1 b = 100 fm^2 = $(10^{-12}$ cm$)^2$ = 10^{-28} m^2
speed (knot)	kn	1 kn = (1852/3600) m/s
logarithmic ratio quantity (neper)	Np	The numerical values of the neper, bel, and decibel (and hence the relation of the bel and the decibel to the neper) are rarely required. They depend on the way in which the logarithmic quantities are defined.
logarithmic ratio quantity (bel)	B	
logarithmic ratio quantity (decibel)	dB	

1.6 SI PREFIXES

Since the presentation of the results in the initial units of physical quantities is sometimes unsuitable, the SI system defines the use of SI prefixes (Table 1.3). The prefixes hecto, deka, deci, and centi are not used in electrical engineering. The names of SI units form the name of the starting unit by adding SI prefixes. Thus, it is correct to say 10 teraohms, but not 10 deciohms. Using only a single prefix in the SI system is also allowed. Therefore, 10 μμΩ is not allowed, as it must be 10 pΩ.

Table 1.3 SI prefixes

Factor	Name	Symbol	Factor	Name	Symbol
10^{24}	yotta	Y	10^{-1}	deci	d
10^{21}	zetta	Z	10^{-2}	centi	c
10^{18}	exa	E	10^{-3}	milli	m
10^{15}	peta	P	10^{-6}	micro	μ
10^{12}	tera	T	10^{-9}	nano	n
10^{9}	giga	G	10^{-12}	pico	p
10^{6}	mega	M	10^{-15}	femto	f
10^{3}	kilo	k	10^{-18}	atto	a
10^{2}	hecto	h	10^{-21}	zepto	z
10^{1}	deka	da	10^{-24}	jocto	y

1.7 BINARY UNITS

The unit **bit** is the shortest information that a computer can work with. It comes from the words **binary** and **digit** and can have only two values – zero and one. Many electronic components can operate like switches that have only two states (ON/OFF), such as transistors, diodes, etc. These components have been the building blocks of computers. Another common unit for storing information in computer technology is the **byte**. With this unit the confusion starts, as the byte does not contain a sequence of **ten** bits, but rather a sequence of **eight** bits. Thus, in the binary system, the byte can have $2^8=256$ possible combinations to store information (from 00000000_{bin} to 11111111_{bin}). However, when it comes to larger quantities of information, there is a problem in using prefixes as defined by SI (see chapter 1.5.). For example, the SI prefix **kilo** has a factor of 10^3, but one **kilobyte** does not have 1000 bytes, but rather 1024 bytes or 2^{10} bytes. This value is close to 1000, but things get worse when larger prefixes are used, such as **mega**, **giga**, or **tera**, which have recently become popular to represent hard disk capacities. Something had to be done, and in 1998, the International Electrotechnical Commission (IEC) attempted to resolve this problem by introducing new prefixes for binary multiples. The prefixes are made in such a way as to take the first two letters of the SI prefixes and add the letters "bi", which stands for the word binary, at the end. Thus, **mega** in the binary world becomes **mebi**, which clearly indicates that it is the 1024 multiple, and not the 1000 multiple. In 1999, IEC published the standard on names and symbols of binary units: "Amendment 2(1999) to IEC Standard 60027-2: Letter symbols to be used in electrical technology, Part 2: Telecommunications and electronics." The multiple of the binary unit N is determined as $N = 2^r$, where **r** can take following values:

$r = 10, 20, 30, 40, 50, 60$

All prefixes for binary units can be seen in Table 1.4., while names, symbols, and values for information systems unit bits and bytes are given in Tables 1.5 and 1.6.

Table 1.4 Binary prefixes

Factor	Name	Symbol	Value of binary prefix
2^{10}	kibi	Ki	$2^{10}=1,024 \cdot 10^3$
2^{20}	mebi	Mi	$2^{20}=1,048576 \cdot 10^6$
2^{30}	gibi	Gi	$2^{30}=1,073\,741\,824 \cdot 10^9$
2^{40}	tebi	Ti	$2^{40}=1,099\,511\,627\,776 \cdot 10^{12}$
2^{50}	pebi	Pi	$2^{50}=1,125\,899\,906\,842\,624 \cdot 10^{15}$
2^{60}	exbi	Ei	$2^{60}=1,152\,921\,504\,606\,846\,976 \cdot 10^{18}$

Table 1.5 Names, symbols and values for unit bit

Name	Symbol	Value
Formative unit bit (symbol: bit)		
kibibit	Kibit	2^{10} bit = 1024 bit
mebibit	Mibit	2^{20} bit = 1024 Kibit
gibibit	Gibit	2^{30} bit = 1024 Mibit
tebibit	Tibit	2^{40} bit = 1024 Gibit
pebibit	Pibit	2^{50} bit = 1024 Tibit
exbibit	Eibit	2^{60} bit = 1024 Pibit

Table 1.6 Names, symbols and values for unit byte

Name	Symbol	Value
Formative unit byte (symbol: B)		
kibibyte	KiB	2^{10} byte = 1024 Bt
mebibyte	MiB	2^{20} byte = 1024 KiB
gibibyte	GiB	2^{30} byte = 1024 MiB
tebibyte	TiB	2^{40} byte = 1024 GiB
pebibyte	PiB	2^{50} byte = 1024 TiB
exbibyte	EiB	2^{60} byte = 1024 PiB

However, the confusion is not over yet, as certain units in information and computer technologies are still decimal. For example, the clock rate of CPU is stated in decimal prefixes, such as 1 GHz CPU; in addition, a 128 kbit/s MP3 stream sends 128,000 bits every second. Hard disks are also stated in decimal units (probably for marketing reasons); therefore, confusion arises after the disk is formatted and the operating system reports less disk space than labeled. Furthermore, CDs are labeled using binary units and DVDs using decimal units, so a 4.7 GB DVD has a binary capacity of only about 4.380,000,000 binary bytes.

Selected bibliography:

1. International Electrotechnical Commission, IEC 60617 - *Graphical Symbols for Diagrams*, 1996
2. *The NIST Reference on Constants, Units and Uncertainty*, http://physics.nist.gov/cuu/Units/index.html, Acquired, August 31st, 2010
3. *Computer Knowledge, Bits, Bytes and Multiple Bytes*, http://www.cknow.com/cms/ref/bits-bytes-and-multiple-bytes.html, Acquired August 31st, 2010.
4. *Mars Climate Orbiter*, http://en.wikipedia.org/wiki/Mars_Climate_Orbiter, Acquired August 31st, 2010.
5. *The Metre Convention*, http://www.bipm.org/en/convention/, Acquired September 06, 2010.
6. J. Q. Shields, R. F. Dziuba, H. P. Layer: "*New Realization of the Ohm and Farad Using the NBS Calculable Capacitor*", IEEE Trans. Instrum. Meas., Vol. 38, No. 2, pp. 249-251, April 1989.
7. M. Reedtz, M.E. Cage: "*An Automated Potenciometric System for Precision Measurement of the Quantized Hall Resistance*", Journal Res. Nat. Bur. Stand., Vol. 99, pp. 303-310.
8. J. Butorac, R. Malarić, I. Leniček: "*Establishment of Resistance Traceability Chain for Croatian Resistance Standards*", Conference Digest of CPEM 2000, Sydney, Australia, pp. 94-95.
9. G. Boella, P. Capra, C. Cassiago i dr.: "*Traceability of the 10 kΩ Standard at IEN*", Conference Digest of CPEM 2000, Sydney, Australia, pp. 98-99.
10. *When is a Kilobyte a Kibibyte? And an MB an MiB?*, http://www.iec.ch/zone/si/si_bytes.htm, Retrieved September 06, 2010.

2. MEASUREMENT ERRORS

Measurements play an important role in modern society, and everyone witnesses a number of measurements every day. Technical sciences are based on data obtained by measurements of various kinds, and not only of electrical quantities. Each measurement is trying to determine the **true value** of the measurand x_p. Through perfect measurement, we believe that we can get a true value. However, using even the most accurate devices and methods, there will always be a deviation between the **measured value** x_i and **true value** x_p. This is due to different causes, such as imperfect measuring equipment, the measurement procedure, the measuring object, and measurer errors. Instead of a "true" value, which is never known, a value called **agreed true value** is used instead. This is the value accepted by agreement, which replaces the true value for a particular purpose. Even when selecting the proper measurement procedure, selecting precise measuring devices, and giving proper attention during the measurement, the measured value can be only more or less close to the true value. The difference between measured values and true values, expressed in units of the measured quantity, is called the **absolute error of measurement Δx**.

One should distinguish between error of **measures** (measuring equipment that embodies a particular value of some size, for example, resistance standards, calibrators, weights, etc.) and errors of the measuring **instruments**. The absolute error of instruments is the difference between the measured value and the true value:

$$\Delta x = x_i - x_p \qquad [2.1]$$

while the absolute error of measures is the difference between the specified (nominal) value x_n and its true value:

$$\Delta x = x_n - x_p \qquad [2.2]$$

To assess the accuracy of measurements, the **relative error of measurement** p is used, and it is the relationship between absolute error and the true value of the measurand:

$$p = \frac{\Delta x}{x_p}. \qquad [2.3]$$

Example 1: The voltmeter measures the voltage $x_i = 1.41$ V, and the true value of voltage is $x_p = 1.44$ V. The absolute error of measurement is $\Delta x = x_i - x_p = -0.03$ V, and the relative error $p = \dfrac{\Delta x}{x_p} = -0.03/1.44 = -0.0208$ or -2.08%.

Example 2: The capacitor's nominal value is 100 pF, and the measured value is 100.1 pF. Its absolute error is then $\Delta x = -0.1$ pF, and the relative error is $p = -0.001$ or -0.1%.

The relative error is expressed in percentages, by multiplying the relative error with 100. For most precise measurements, the relative error is expressed in parts per million - ppm (1 ppm = 10–6).

The concept of **correction** is also important. It has the same absolute value as the absolute error, but the opposite sign:

$$k = -\Delta x \qquad [2.4]$$

Knowing the absolute error, by using correction, the true value can be determined:

$$x_p = x_i + k \qquad [2.5]$$

This principle applies only in theory, because the correction too has some uncertainty, so the true value can only be more or less approached.

Errors are divided into three types: grave, systematic, and random.

2.1 GRAVE ERRORS

Grave errors are caused by negligence of the operator, the choice of inadequate equipment, or the measurement method. An example of a grave error is the wrong reading of results, e.g., reading 23.8 instead of 28.3 on a digital instrument. Grave errors also occur due to incompetence of the measurers, or due to incorrect calculations. Since grave errors remain unnoticed in most cases, such errors cannot be mathematically evaluated, or taken into account through a correction. Grave errors are avoided with good knowledge and attention during the measurements, and a proper selection of measurement equipment and the measurement procedure. It is always useful to evaluate the approximate value of the measurand before the measurements.

2.2 SYSTEMATIC ERRORS

Systematic errors generally occur because of imperfections of the scale, measure, measuring object, and measurement methods. For example, systematic errors are caused by connected instruments in cases where they need little power to function, voltmeter internal impedance can cause errors in resistance measurements, and ammeters cause systematic errors in electrical current measurements. Systematic errors also arise due to measurable environmental influences (humidity, pressure, temperature, magnetic field, etc.). An important characteristic of systematic errors is that they have a constant value and sign, and these can be taken into account when correcting the measurement result.

2.3 RANDOM ERRORS

Random errors result from small and variant changes that occur in the standards, measures, measurement laboratory, and environment. These errors can cause a large number of different errors in each individual measurement and each time have a different size and sign, causing measurement results to scatter.

If the same quantity is measured several times in a row under the same external conditions and with the same instrument, each time the results will scatter around some value due to random errors that change the size and sign in each measurement. These random errors cannot be solved by making corrections, as for some systematic errors. Since all measurements are carried out under the same conditions, all results have the same weight. Thus, the most probable value of the measured quantity is the arithmetic mean of individual results. If the measurements are repeated n times, and individual results are $x_1 + x_2 + x_3, ..., x_n$ then the arithmetic mean of the individual results $\bar{x}$ is:

$$\bar{x} = \frac{x_1 + x_2 + x_3 + ... + x_n}{n} = \sum_{i=1}^{n} x_i \, .\qquad [2.6]$$

The measurement is more precise if the measurement results scatter less from one another. The measure of precision is **standard deviation** (mean square error) of individual measurements according to the following equation:

$$s = \sqrt{\frac{1}{n-1} \sum_{i=1}^{n} \left(x_i - \bar{x}\right)^2}\qquad [2.7]$$

According to the above expression the **mean square error** or **standard deviation** is calculated on the basis of individual differences between individual results x_i and the mean value of measurements $\bar{x}$, rather than the differences between individual results and true value, which would correspond to the definition of error. The reason for this is that the true value in most cases is not known, and the mean square error can only be approximated by statistical tools. Standard deviation can be calculated in relative value by dividing it with the arithmetic mean.

The arithmetic mean of all measurements is usually taken as a result of measurement, so we need to know how much the deviation of the arithmetic mean amounts to. The **standard deviation of arithmetic mean** is calculated by the following equation:

$$s_{\bar{x}} = \frac{s}{\sqrt{n}} \, .\qquad [2.8]$$

The resulting standard deviation of the arithmetic mean is less than the standard deviation of individual measurements and is inversely proportional to the square root of the number of measurements; therefore, performing more measurements to reduce the deviation makes sense only to a certain extent. For example, a measurement that was repeated 100 times will have the standard deviation of approximately 3.16 times less than the standard deviation of measurements repeated 10 times, while the measurement that was repeated 1000 times will have only a ten times smaller deviation.

Smaller standard deviation means less dispersion of results and better **precision**, but it should not be confused with the term **accuracy**. Precision is associated with the degree of repeatability and closeness of individual results obtained by repeated measurements of some quantity under the same conditions.

Measurement repeatability is the closeness of the individual repeated measurements of the same quantity under the same conditions using the same equipment.

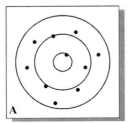 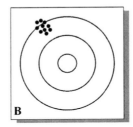 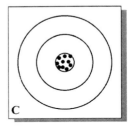

Figure 2.1 Precision and accuracy

Precision is an essential requirement for accuracy, but it is not sufficient. **Accuracy** indicates the closeness of the results to the true value. Accuracy, on the other, hand ensures precision.

The example that best illustrates the difference between **precision** and **accuracy** is shooting at a target (Figure 2.1). The first shooter (A) is **imprecise** and **inaccurate** because his hits are scattered and far from the center of the target. The second shooter is **precise**, but **inaccurate**, because his shootings disperse a little, but are far from the target center, so he has "systematic error." The third shooter (C) is **precise** and **accurate** because his shootings are a little dispersed, but are also positioned in the center of the target.

Good matching of individual measurements is not in itself proof that the result is correct. Therefore, it is necessary to define another term - reproducibility of results.

Measurement reproducibility is the closeness of individual measurements of the same quantity that are measured in changed circumstances (different measurers in different laboratories, using other measurement methods, instruments, places, and conditions, etc.).

There are also two other terms usually associated with the terms accuracy and precision: **sensitivity** and **resolution**.

Sensitivity is the degree of response of a measuring instrument to the change of the measured quantity Δx which has caused this change. Resolution is the smallest change in measured values that the measuring instrument can register. It is important to note that a claim by a manufacturer that his device has great resolution and high sensitivity is no guarantee that the device is also accurate.

If all measurements are not carried out under the same conditions, all results will not have the same weight, and the standard deviation of one set of measurements will differ from the other set of measurements. In this case, the most probable value of the measurand is determined using the **general mean**:

$$\overline{x} = \frac{p_1\overline{x_1} + p_2\overline{x_2} + \ldots + p_m\overline{x_m}}{p_1 + p_2 + \ldots + p_m},$$ [2.9]

where $\overline{x_1}$, $\overline{x_2}$, ..., $\overline{x_m}$ are the arithmetic mean of m individual sets of measurements, and $p_1, p_2, \ldots, p_m$ are weights of each set of measurements. Weights are calculated according to:

$$p_i = \frac{K}{s_{\overline{x_i}}^2}.$$ [2.10]

The constant K is arbitrarily chosen to make the calculation easier. The standard deviation of a general mean is determined by the formula:

$$s_{\overline{x_s}} = \frac{1}{\sqrt{\sum_{i=1}^{n} \frac{1}{s_{\overline{x_i}}^2}}}.$$ [2.11]

If errors occur due to a large number of random and mutually independent causes, each of which causes various errors, the results are then subject to the Gaussian or normal distribution. Normal distribution was mathematically elaborated by German mathematician **Karl Gauss** in the year 1808, and independently by the American scientist **Robert Adrian**. However, it was the Italian astronomer **Galileo Galilei** who in 1632 published several postulates about the random errors in measurement:

1. that they are inevitable
2. that smaller errors are more frequent than large errors
3. that errors equal in magnitude, but with the opposite sign, are equally likely
4. that the true value is in the highest concentration of measurement results.

Today, many problems in various application areas are solved by using this distribution. Weights and heights of people, water consumption in the city during the day, distance from the center of the target, the number of patients admitted to a hospital during the day, the size of an oak's leaves, and numerous other examples are subject to Gaussian distribution.

The Gaussian or normal distribution probability function is defined as:

$$f(x) = \frac{1}{\sigma\sqrt{2\pi}} e^{-\frac{1}{2}\left(\frac{x-\overline{x_0}}{\sigma}\right)^2}$$ [2.12]

where $\overline{x_0}$ is arithmetic mean and σ is standard deviation of infinite number of measurements. The arithmetic mean and standard deviation of Gaussian distribution (Figure 2.2) is unambiguously defined. The area under the curve of normal distribution, $\infty < x < \infty$ is equal to one, i.e., the probability that the value of the measurand lies within these limits is 100%. With the presence of random errors it is not possible to determine the true value of the measurand, but only limits of the area within which we can expect the true value of measurand with a certain probability.

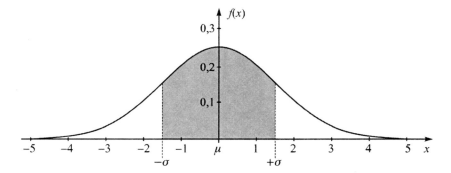

Figure 2.2 Gaussian or normal distribution

The probability $P_{(x1<x>x2)}$ that the measured value **x** will fall between x_1 and x_2 is obtained by integrating the probability function within the limits of the x_1 to x_2.

$$P(x_1 < x < x_2) = \frac{1}{\sigma\sqrt{2\pi}} \int_{x_1}^{x_2} e^{-\frac{1}{2}\left(\frac{x-\overline{x_0}}{\sigma}\right)^2} dx . \qquad [2.13]$$

This integral actually represents an area under the curve of probability of interval from x1 to x_2. Some of the most often used probabilities are given in Table 2.1.

Table 2.1. The probability that the true value lies within certain limits

Lower and upper bounds	The probability that x is	
	Within the limits	Outside the limits
$\overline{x_0} \pm 0{,}674\sigma$	50%	50%
$\overline{x_0} \pm \sigma$	68.26%	31.74%
$\overline{x_0} \pm 2\sigma$	95.45%	4.55%
$\overline{x_0} \pm 3\sigma$	99.73%	0-27%

For Gaussian type measurements 68.3 % of all possible results lie between the limits $-\sigma$ i $+\sigma$, while between the limits -2σ i $+2\sigma$ there are already 95% of all results.

2.4 CONFIDENCE LIMITS

In most cases the arithmetic mean is not equal to the true value of the measurand, even if all systematic errors are removed. Only the arithmetic mean of an infinite set of measurements becomes a true value. Therefore, bounds are defined in which, with some probability P, a true value can be expected. These limits are called **confidence limits,** and the area $\overline{x_0} \pm \dfrac{k\sigma}{\sqrt{n}}$ within those limits is the **area of**

MEASUREMENT ERRORS

confidence. The standard deviation of the infinite number of measurements σ is unknown, but if the measurements are repeated a sufficient number of times (n>30), then the standard deviation s can be considered a good estimate of the standard deviation σ. Table 2.2 shows the area of confidence for the probabilities of 68.3%, 95% and 99%.

Table 2.2. Confidence limits

Area under the Gauss curve	Confidence intervals	Confidence limits	
		lower	upper
P=68.3 %	$\overline{x_0} \pm \dfrac{\sigma}{\sqrt{n}}$	$\overline{x_0} - \dfrac{\sigma}{\sqrt{n}}$	$\overline{x_0} + \dfrac{\sigma}{\sqrt{n}}$
P=95 %	$\overline{x_0} \pm \dfrac{1{,}96\sigma}{\sqrt{n}}$	$\overline{x_0} - \dfrac{1{,}96\sigma}{\sqrt{n}}$	$\overline{x_0} + \dfrac{1{,}96\sigma}{\sqrt{n}}$
P=99%	$\overline{x_0} \pm \dfrac{2{,}58\sigma}{\sqrt{n}}$	$\overline{x_0} - \dfrac{2{,}58\sigma}{\sqrt{n}}$	$\overline{x_0} + \dfrac{2{,}58\sigma}{\sqrt{n}}$

Example 3: The output of a voltage source was measured 10 times. The mean value of measured voltage was $\overline{x}$ = 3.4 V and the standard deviation of individual measurements was s = 0.1 V. Standard deviation of the mean can be calculated $s_{\overline{x}} = \dfrac{s}{\sqrt{n}}$ = 0.032 V, and with the probability of 68.3% it can be said that the measured value lies in the following limits $\overline{x} \pm \dfrac{s}{\sqrt{n}}$ = (3,4±0,032) V, and with the probability of 99% it can be said that the measured value lies within limits $\overline{x} \pm \dfrac{2{,}58 s}{\sqrt{n}}$ =(3,4±0,081) V.

In the case of only a few individual measurements, the confidence interval is determined by **student's t-distribution**:

$$\overline{x} \pm \dfrac{t}{\sqrt{n}} s . \qquad [2.14]$$

Factor **t** depends on probability P and the number of individual measurements n. Some values for $\dfrac{t}{\sqrt{n}}$ are given in Table 2.3.

Table 2.3. Values for $\dfrac{t}{\sqrt{n}}$

n \ P	3	5	10	30	50	100
68,3 %	0,76	0,51	0,34	0,19	0,14	0,10
95 %	2,5	1,24	0,72	0,37	0,28	0,20
99 %	5,7	2,1	1,03	0,50	0,38	0,26

According to the student's distribution for the above example with the probability of 68.3%, it can be said that the measured value lies within the following interval $\bar{x} \pm \dfrac{t}{\sqrt{n}} s$ =(3.4±0.034) V.

2.5 MEASUREMENT UNCERTAINTY

The determination and expression of the uncertainty of measurement has been a subject of debate in a number of metrological organizations (IEC, BIPM, ISO, etc.) around the world for many years. A number of recommendations, guidelines, and instructions have been generated. The latest internationally-accepted document for the expression of measurement uncertainty is **EA-4/02**, which was adopted by the European Cooperation for Accreditation (EA). **Measurement uncertainty** is defined as "a parameter, associated with the result of a measurement that characterises the dispersion of the values that could reasonably be attributed to the measurand."

The confidence limits described in Chapter 2.4 take into account only random errors, and assume that all causes of systematic error are either removed or corrected. Since it is virtually impossible to achieve this in most cases, the measurement uncertainty must include all the remaining systematic errors as the second component of uncertainty. The share of errors in measurement uncertainties that are treated with statistical methods is marked with u_A (type A standard uncertainty), while the remaining errors are identified by u_B (type B standard uncertainty). **Standard uncertainty of type A** is obtained from the mean value of the measurand, whereas **type B standard uncertainty** is based on:

a) preliminary results of measurement,
b) experience or general knowledge about the behavior and characteristics of the measuring instrument,
c) manufacturer's data,
d) calibration data, and
e) uncertainties of individual constants specified in the manuals.

Measurement uncertainty is then calculated according to the formula:

$$u = \sqrt{u_A^2 + u_B^2} \ . \qquad [2.15]$$

In the case of multiple components of systematic error type B, standard uncertainty is calculated by:

$$u_B = \sqrt{\sum_{i=1}^{n} u_i^2} \ . \qquad [2.16]$$

The result of measurement is then expressed as:

$$\bar{x} \pm u \ . \qquad [2.17]$$

Measurement uncertainty can also be expressed in relative form; the result is then expressed by:

$$\overline{x}\left(1 \pm \frac{u}{\overline{x}}\right). \qquad [2.18]$$

By increasing the number of measurements, only the standard uncertainty of type **A** can be reduced, while the standard uncertainty of type **B** remains unchanged, so it is unnecessary to repeat the measurement many times. When expressing uncertainty, it is necessary to provide information about each component of uncertainty, the manner and methods for their calculation, and all the corrections of systematic errors.

2.6 LIMITS OF ERROR (SPECIFICATIONS)

Limits of error define the maximum deviation of the measurement device output from the true value. They must include all systematic errors and the effect of equipment aging. Limits of error allow for the easy division of devices into working and malfunctioning categories. It is therefore necessary that the measurement uncertainty for a device is small enough to avoid the dubious situation in assessing the status of devices; preferably, it does not exceed ¼ of the stated error limits of the device. Limits of error should be distinguished from the confidence limits and measurement uncertainty.

Example 4: Resistors with a nominal value of 10 Ω have guaranteed limits of error ±0.05 Ω. Three different resistors were compared with the standard resistor and the following resistor values of resistors were obtained (Figure 2.3).

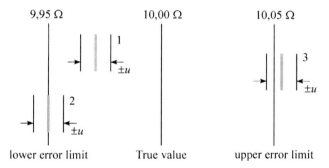

Figure 2.3 The difference between the limits of error and measurement uncertainty

For the first resistor the value obtained is 9,97 ± 0,01 Ω; for the second, the value is 9,95 ± 0,01 Ω; and the third value is 10,06 ± 0,01 Ω. The first resistor is correct without doubt, and the limits of errors also hold for the second resistor. The third resistor is, however, defective, because its measured value is outside the guaranteed limits of error. All measured values within the error limits have the same probability, and results outside these limits are not possible.

Such a distribution of measurement results is called **rectangular distribution** (Figure 2.4), and the standard deviation of individual results is $\dfrac{a}{\sqrt{3}}$.

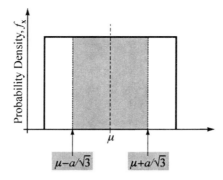

Figure 2.4 Rectangular distribution

2.6.1 Indirectly Measured Quantities

Often a value is not measured directly but is obtained through a calculation based on measurements of other quantities. For example, the resistance of a resistor is determined by measuring the voltage drop on it and the current that runs through it. What, then, is the standard deviation and limit of error from the results obtained?

If the measured quantity y is the function of several independent, directly-measured quantities:

$$y = F(x_1, x_2, ..., x_n) \qquad [2.19]$$

and each of these quantities has a standard deviation due to the effects of random errors, then the standard deviation of y is determined by the following equation:

$$s_y = \sqrt{\sum_{i=1}^{n}\left(\frac{\partial F}{\partial x_i} s_i\right)^2}. \qquad [2.20]$$

Some of the simpler functions have the standard deviation:

a) the sum of: $y = x_1 + x_2$:

$$s_y = \sqrt{s_1^2 + s_2^2} \quad \text{and} \quad s_{y\%} = \frac{\sqrt{x_1^2 s_{1\%}^2 + x_2^2 s_{2\%}^2}}{x_1 + x_2} \qquad [2.21]$$

b) the difference: $y = x_1 - x_2$:

$$s_y = \sqrt{s_1^2 + s_2^2} \quad \text{and} \quad s_{y\%} = \frac{\sqrt{x_1^2 s_{1\%}^2 + x_2^2 s_{2\%}^2}}{x_1 - x_2} \qquad [2.22]$$

The standard deviation of the difference in percentages can become very large, although the standard deviation of measured values x_1 and x_2 is very small. This occurs when the difference between x_1 and x_2 is very small - such measurements should be avoided.

c) The product: $y = x_1 \cdot x_2$:

$$s_y = \sqrt{x_2^2 s_1^2 + x_1^2 s_2^2} \text{ and } s_{y\%} = \sqrt{s_{1\%}^2 + s_{2\%}^2} \quad [2.23]$$

d) quotient: $y = \dfrac{x_1}{x_2}$:

$$s_y = \sqrt{\dfrac{s_1^2}{x_2^2} + \dfrac{x_1^2 s_2^2}{x_2^4}} \text{ and } s_{y\%} = \sqrt{s_{1\%}^2 + s_{2\%}^2}. \quad [2.24]$$

In practice, only the limits of errors of measurement devices used in measurements are known. If the result is calculated on the basis of measurements of some other quantities, then the **safe limits of error** are determined by the following equation:

$$G'_y = \pm \sum_{i=1}^{n} \left| \dfrac{\partial F}{\partial x_i} G_i \right|. \quad [2.25]$$

The limits of error will not be exceeded for certain only if they are determined by summing the absolute values of partial derivations of the function.

Safe limits of errors of some simple functions:

e) the sum of: $y = x_1 + x_2$:

$$G'_y = \pm \{|G_1| + |G_2|\} \text{ and } G'_{y\%} = \pm \dfrac{|x_1 G_{1\%}| + |x_2 G_{2\%}|}{x_1 + x_2} \quad [2.26]$$

f) the difference: $y = x_1 - x_2$:

$$G'_y = \pm \{|G_1| + |G_2|\} \text{ and } G'_{y\%} = \pm \dfrac{|x_1 G_{1\%}| + |x_2 G_{2\%}|}{x_1 - x_2} \quad [2.27]$$

g) the product: $y = x_1 \cdot x_2$:

$$G'_y = \pm \{|x_2 G_1| + |x_1 G_2|\} \text{ and } G'_{y\%} = \pm \{|G_{1\%}| + |G_{2\%}|\} \quad [2.28]$$

h) the quotient: $y = \dfrac{x_1}{x_2}$:

$$G'_y = \pm\left\{\left|\dfrac{G_1}{x_2}\right| + \left|\dfrac{x_1 G_2}{x_2^2}\right|\right\} \text{ and } G'_{y\%} = \pm\left\{|G_{1\%}| + |G_{2\%}|\right\} \quad [2.29]$$

With the increase of required quantities for determining the function's limits of error, it is less likely that the results error will reach these calculated limits, so in practice **statistical error limits** are frequently used. They are calculated according to the following equation:

$$G'_y = \sqrt{\sum_{i=1}^{n}\left(\dfrac{\partial F}{\partial x_i} G_i\right)^2} . \quad [2.30]$$

The term is identical to the expression for the standard deviation of the indirectly measured quantity y, except that instead of the standard deviation, error limits of measurement quantities are used.

Selected bibliography:
1. B. N. Taylor, C. E. Kuyatt: *"Guidelines for Evaluating and Expressing the Uncertainty of NIST Measurement Results"*, NIST Technical Note 1297, 1994 Edition.
2. The NIST Reference on Constants, Units, and Uncertainty, *"Uncertainty of Measurements Results"*, http://physics.nist.gov/cuu/Uncertainty/index.html, Acquired September 06, 2010.
3. V. Bego, *"Measurements in Electrical Engineering"*, (in Croatian) Tehnička knjiga, Zagreb, 1990
4. V. Bego: *"Errors in Electrical Engineering Measurements"*, (in Croatian), Department of Electrical Engineering and Measurements, Faculty of Electrical Engineering and Computing, Zagreb, 1985
5. L. Feil, *"The Theory of Errors and the Adjustment Calculation"*, (in Croatian), Faculty of Geodesy, Zagreb, 1990
6. N. Čubranić, *"The Theory of Errors and the Adjustment Calculation"*, (in Croatian), Tehnička knjiga, Zagreb, 1966
7. International Organization for Standardization, *"Guide to the Expression of Uncertainty in Measurement"*, 1993.
8. A. Singmin: *"Type A/Type B Uncertainty Calculation Demystified"*, Cal Lab, pp. 32-38, September-October 1997.
9. European Cooperation for Accreditation, EA-4/02, *"Expression of the Uncertainty of Measurements in Calibration"*, 1999
10. P. Crisp: *"Uncertainty Analysis for Laboratory Accreditation"*, Wavetek Calibration Instruments Division.,
11. http://assets.fluke.com/appnotes/Calibration/Uncertainty_Analysis_for_Laboratory_Accreditation.pdf, Acquired September 06, 2010.

3. MEASURING ELEMENTS

3.1 RESISTORS

A resistor is an electronic component that produces a voltage across its terminals that is proportional to the electric current passing through it, in accordance with Ohm's law (R=U/I). As resistors are the most common elements in electrical circuits that have a number of different purposes, there is a wide spectrum of resistor types on the market today. Some of them are made of various materials and films, and resistance wires. However, all of the resistors have some common characteristics. The most important characteristic is the resistance, but also important are the temperature coefficient, power rating, noise, self-inductance, and capacitance. Each of these characteristics can be important in certain applications, so a good knowledge of them is needed to achieve the required accuracy and secure operation.

3.1.1 Equivalent Circuit of the Resistor

In addition to active resistance, the resistor also has its self-inductance, while between the coils, casings, and bodies, a parasitic resistor's self-capacitance occurs. These capacitances and inductances are manifested in alternating current circuits, and resistors are then displayed in an **equivalent circuit** as impedance consisting of resistance R, inductance L, and the capacitance C (Figure 3.1).

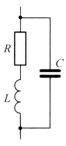

Figure 3.1 Equivalent circuit of the resistor

Self-capacitance is particularly evident at high frequencies. At low resistance, inductance effects dominate, and are usually reduced by special winding (**bifilar**) (Figure 3.2), while capacitance effects are reduced by dividing winding in several sections.

Figure 3.2 Resistor bifilar winding

The quality measurement resistor in AC circuits is expressed by time constant τ:

$$\tau = \frac{L}{R}\left(1 - \omega^2 LC\right) - RC. \qquad [3.1]$$

At frequencies below 20 kHz, $\omega LC \ll 1$ is valid, and the time constant can be approximated:

$$\tau = \frac{L}{R} - RC \qquad [3.2]$$

If $\frac{L}{R} = RC$, the time constant will be equal to zero, even though the measuring resistor has a certain self-inductance and capacitance. High quality resistors usually have a small time constant which is expressed in nanoseconds (ns).

In addition to wire resistors, **layer resistors** are also used. In these, a thin layer of metal, metal oxides, and carbon is applied to the body of the insulator (porcelain, glass, ceramics).

3.1.2 Resistance Decades and Slide Resistors

Resistance decade boxes have resistors of many values which are usually decimally graded and located in a common box. For each decade, there is usually a note indicating the maximum power rating (Figure 3.3.).

Figure 3.3 Resistor decade box

The **slide resistor** (variable resistor, potentiometer) allows the continuous change of resistance. In such resistors, the resistance wire is wound around the cylinder, which is made of insulating material. After the cord slider slides in the form of a pen, it allows the continuous change of resistance. The motion of the slider can be linear or circular (Figure 3.4a and 3.4b).

MEASURING ELEMENTS

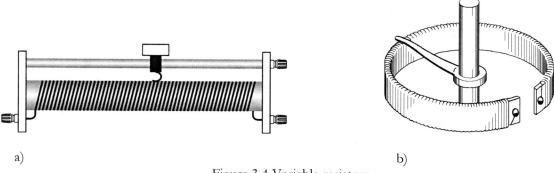

a) b)

Figure 3.4 Variable resistors

3.1.3 Resistance Standards

For the entire 20th century, before the discovery of the quantum Hall effect, the unit Ohm was stored by using the resistance standards made from resistive wire, usually manganin. The first manganin standards were made in the German institute **Physikalisch Technische Reichsanstalt** (PTR) in 1892. Today this type of standards is still used, usually for standards of small nominal resistance value of 1 mΩ to 100 mΩ, where large currents are needed for calibration. All standards of resistance for a small nominal value are made with four terminals, two current and two voltage terminals, while higher nominal value standards are made with only two terminals. The resistance standard **R** with four terminals is defined as the ratio of voltage between the voltage terminals **P1** and **P2** and the current passing through the current terminals **S1** and **S2** (Figure 3.5).

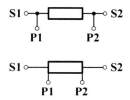

Figure 3.5 Voltage terminals **P1** and **P2** and current terminals S1 and S2

Standards of lower nominal values are made with four terminals because the resistance of copper wires and contact resistance on the terminals cannot be ignored. Sometimes these resistances amount to more than 100 mΩ, which is larger than the resistances to be measured. In such a case, not even small temperature changes of copper wires can be ignored. Therefore, these resistors have voltage terminals that are attached directly to the end of the manganin wire. The voltage drops on the voltage terminals have no influence on the measurement result because they have negligible voltage levels, due to small currents in the voltage measurement circuit. The current in the voltage measurement circuit depends on the internal resistance of the voltmeter, but it is usually less than 10^{-9} A in high precision measurement methods, as can be seen in Figure 3.6.

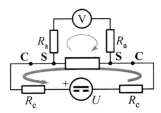

Figure 3.6 Current and sense voltmeter circuit in four terminal resistance standards (the bold line indicates large current - usually several amperes for small resistances - and the thin line indicates small current)

In Figure 3.7, several types of standard resistors are shown with four terminals. The first resistor on the left is supposed to be used in air baths or in laboratory ambient conditions, while the rest are designed to be used in oil baths. These standards are included in the group resistance standards in the Primary Electromagnetic Laboratory of Croatia.

Figure 3.7 Resistance standards of different types in the Primary Electromagnetic Laboratory of Croatia (from left to right: Fluke, VEB, Guildline, Leeds&Northrup and Siemens&Halske)

The construction of a typical standard resistor designed to be used in oil baths is presented in Figure 3.8, where the resistive manganin wire is wound on the internal cylinder of the resistor. The thermometer can be placed in the middle of the resistor to control the temperature of the standard during the measurement process.

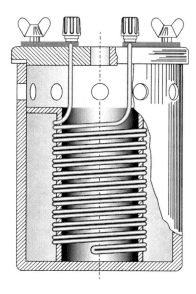

Figure 3.8 Internal view of a standard resistor, with the wires in bifilar coil arrangement

3.1.4 Oil Bath

Standards are mostly stored in thermostated oil baths; the temperature of most of these thermostats is kept at 23 °C. The maximum resistor power rating in the air is usually lower than 1 W and 10 W in oil baths, although the most accurate measurement standard does not exceed more than 100 mW, or even 10 mW, as some of them exhibit permanent change of value (hysteresis effect) after they are exposed to higher power dissipation. The oil bath UTEO-94 designed at the Primary Electromagnetic Laboratory of Croatia is shown in Figure 3.9; the top view of the oil bath is seen in Figure 3.10. The oil bath UTEO-94 is a double cylinder bath filled with silicon oil, and the temperature is maintained at 23 °C with a regulator that controls the Peltier cooler and electric heater.

The heater is placed at the bottom of the bath, while Peltier coolers are placed at the top, which causes the oil to flow from the bottom to the top of the internal cylinder, where the oil overflows and passes through the narrow gap down, cooled by Peltier coolers. The temperature gradient is not more than several millikelvins from top to bottom. The bath is also temperature-insulated from outside effects; it can store up to twelve standard resistors and has also a place for one external resistor for calibration purposes.

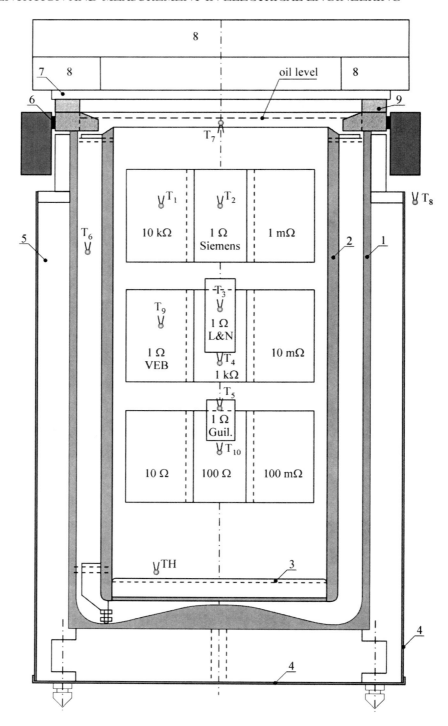

Figure 3.9 Oil bath with thermostat for maintaining standard resistors of the Primary Electromagnetic Laboratory of Croatia at constant temperature – side view (1- outer container; 2 – inner container; 3 – heating coil; 4 – casing; 5 – thermal insulation; 6 – Peltier elements and coolers; 7 - cover with current and voltage terminals of standard resistors; 8 – thermal insulation; 9 – Seal)

MEASURING ELEMENTS

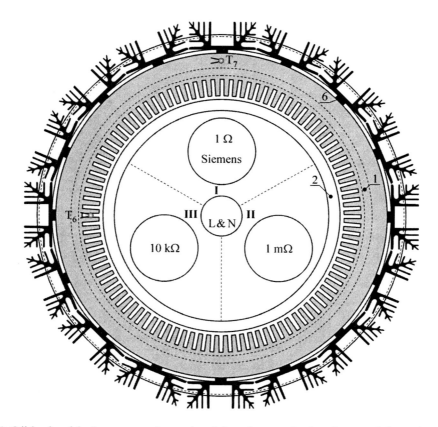

Figure 3.10 Oil bath with thermostat for maintaining the standard resistors of the Primary Electromagnetic Laboratory of Croatia at constant temperature – Top view (1- outer container; 2 – inner container; 6 - 6 – Peltier elements and coolers)

3.1.5 Group Standards

In laboratories that do not have a quantum Hall resistance standard, the primary standard of resistance usually consists of a bank of several standard resistances of 1 Ω, and its mean value is closely monitored until the next calibration with quantum Hall standards. The traceability of Croatian resistance standards can be seen in Figure 3.11. International traceability of the group resistance standards is achieved with the calibration of several standard resistances at the German National Metrology Institute PTB, while all the other standard resistances are calibrated at the Croatian Primary Electromagnetic Laboratory using the comparison methods with nominal ratios 1:1 and 1:10.

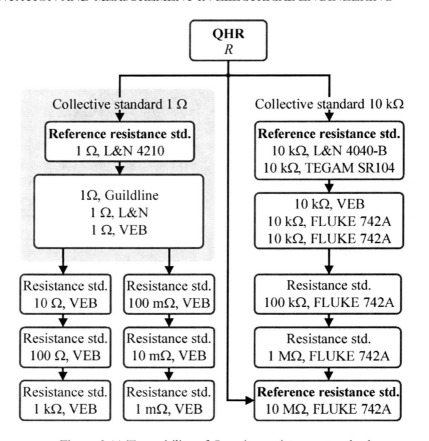

Figure 3.11 Traceability of Croatian resistance standards

Calibrations at international metrology institutes are costly, and every national metrology institute must secure the reliability of its standards in between the international calibrations that are required every two or three years. The reliability is satisfactory when the standards have as little time drift as possible between calibrations. If the drift of each individual standard is unpredictable, such as in Figure 3.12, this can be improved by inter-comparing each of the standard resistances with all the others of the same nominal value.

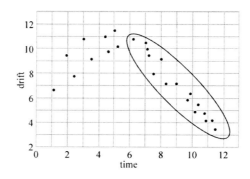

Figure 3.12 Unpredictable drift of standard over-time

If full inter-comparison is complicated for some reason, it can be simplified, as standards can be partially inter-compared in circular motion. Such inter-comparison is called ring comparison (or round robin), and in this case, the first standard is compared with the second, the second with the third, and finally the last standard is compared with the first one. For example, if there are four standards of the same nominal value (as is the case for 1 Ω and 10 kΩ levels at the Croatian Primary Electromagnetic Laboratory - Figure 3.11), and if the measured relative differences between the first and the second, the second and the third, the third and the fourth, and the fourth and the first are $\delta_{1,2}$; $\delta_{2,3}$; $\delta_{3,4}$; and $\delta_{4,1}$ respectively, and if $\delta_1, \delta_2, \delta_3$ and δ_4 are relative deviations of each standard from the nominal value, it can be written:

$$\delta_{1,2} = \delta_1 - \delta_2 \qquad [3.3]$$
$$\delta_{2,3} = \delta_2 - \delta_3 \qquad [3.4]$$
$$\delta_{3,4} = \delta_3 - \delta_4 \qquad [3.5]$$
$$\delta_{4,1} = \delta_4 - \delta_1 \qquad [3.6]$$

In ring comparison, the following is valid:

$$\delta_{1,2} + \delta_{2,3} + \delta_{3,4} + \delta_{4,1} + \omega = 0 \qquad [3.7]$$

If all measurements (comparisons) were perfectly accurate, the sum $\delta_{1,2} + \delta_{2,3} + \delta_{3,4} + \delta_{4,1}$ would be zero, but due to inevitable errors, the sum will be equal to $-\omega$. As all the comparisons have equal weight, the correction of each comparison is equal to:

$$v = \frac{\omega}{N} \qquad [3.8]$$

where N is the number of standards involved in a ring comparison.

In high precision measurement methods (such as presented in Chapter 10), the corrections v are usually smaller than 10^{-8}, which is itself an indication of measurement method accuracy.

The mean value deviation of four standards from the nominal value is:

$$\delta_{sr} = \frac{(\delta_1 + \delta_2 + \delta_3 + \delta_4)}{4} \qquad [3.9]$$

The ring comparison only measures the relative deviation of resistances. If the mean value of the group is known to be δ_{sr}, then each individual standard deviation from the nominal value can be calculated using the ring comparison with the following equations:

$$\delta_1 = \delta_{sr} + \frac{(\delta_{1,2} - \delta_{3,4} - 2 \cdot \delta_{4,1})}{4} \qquad [3.10]$$

$$\delta_2 = \delta_{sr} + \frac{(\delta_{2,3} - \delta_{4,1} - 2 \cdot \delta_{1,2})}{4} \qquad [3.11]$$

$$\delta_3 = \delta_{sr} + \frac{(\delta_{3,4} - \delta_{1,2} - 2 \cdot \delta_{2,3})}{4} \qquad [3.12]$$

$$\delta_4 = \delta_{sr} + \frac{(\delta_{4,1} - \delta_{2,3} - 2 \cdot \delta_{3,4})}{4} \qquad [3.13]$$

The equations above are actually approximations, but with accuracy levels below 10^{-8} if the resistances (or more general standards) do not differ from each other by more than 0.01 % of the same value, which is usually true for the most precise standards used in national metrology institutes. The ring comparison can be used then to track the drift of each standard resistance value in time from the mean value of a group standard. This procedure is useful, as the mean value is expected to drift less than each individual standard resistance value. However, it is necessary to calibrate the mean value of the group standard occasionally (every two or three years). To do this, it is necessary to calibrate only one standard of the group standard in an international (or better equipped) laboratory. If, for example, the first standard resistance is calibrated, then the new mean value will be obtained by the following calculation:

$$\delta_{sr} = \delta_1 - \frac{(\delta_{1,2} - \delta_{3,4} - 2 \cdot \delta_{4,1})}{4} \qquad [3.14]$$

and then all other resistances can be calculated by mutual comparisons (3.8-3.11).

Materials for standard resistors must have good time stability through decades of use, small temperature coefficients, high resistivity, small thermoelectric EMF of copper to manganin, and be resistant to mechanical shock and stress. Alloys of copper, manganese, and nickel have been used for making standard resistors from the moment of their discovery in 1884 up to the present day. The most frequent combination of these metals, *manganin*, contains 84% copper, 12% manganese, and 4% nickel. The manganin-copper thermal EMF is less than 2.2 µV/°C, with a resistivity of 0.48 µΩ·m, and temperature coefficient of less than 10 (µΩ/Ω)/°C. Temperature characteristics of manganin are cubic-shaped, but for more than ten degrees on each side of the maximum, which lies in the 20 °C to 30 °C range, they can be expressed by the equation:

$$R_t = R_0[1 + a(t - t_0) + \beta(t - t_0)^2] \qquad [3.15]$$

where α is the slope of the curve and β its curvature with temperature t_0.

Standard resistors are maintained at various temperatures ranging from from 20 °C to 25 °C in the laboratories, but 23 °C is the most frequent setting. If the temperature coefficients α and β declared by the manufacturer are different from the temperature maintained by the laboratory, it is necessary to recalculate the coefficients to the desired temperature. For example, to change the coefficients from 20 °C (a_{20} and β_{20}) to 23 °C (a_{23} and β_{23}), the following equations can be used:

$$\alpha_{23} = \frac{R_{20}}{R_{23}}(\alpha_{20} + 6\,°C \cdot \beta_{20}) \qquad [3.16]$$

$$\beta_{23} = \beta_{20} \cdot \frac{R_{20}}{R_{23}} \qquad [3.17]$$

where R_{20} and R_{23} are resistance values at 20 °C and 30 °C, respectively.

3.1.6 Hamon Transfer Resistor

If a group of equal value resistors (usually 10 or 100) can be connected both in series and in parallel (possibly by some switch), then it is possible to easily calibrate another ohmic level that is otherwise calibrated with much higher uncertainty. According to the Hamon network, the following is valid:

$$\frac{R_S}{R_P} = n^2 \left(1 + \frac{1}{n} \sum_{i=1}^{n} m_i^2 \right) \qquad [3.18]$$

where R_S is the total resistance of n resistances serially connected, R_P is the total resistance of n resistances parallelly connected, and m_i is the deviation of resistance R_i from the mean value. For example, if the required accuracy of the serial to parallel ratio is 10^{-8}, then the deviation of n individual resistors from the mean value must be less than 10^{-4}. If the required accuracy is 10^{-6} or 1 ppm (part per million), then the deviation of n individual resistors from the mean value must be less than 10^{-3}. Therefore, with the careful selection of resistors, greater accuracy of ratios can be achieved.

Figure 3.13 shows the Hamon resistor VOO-100, designed at the Primary Electromagnetic Laboratory of Croatia, which is made of carefully chosen one hundred metal film resistors of 1 MΩ that make 100 MΩ if connected serially, and 10 kΩ if connected in parallel. Due to careful construction, the isolation resistors are equal in serial and parallel connections, and do not influence the ratio. The Hamon resistor is used as a transfer of the 10 kΩ level to the 100 MΩ level in one step, as the 10 kΩ level has low uncertainty. The laboratory owns another self-made Hamon resistor that has a ten 1 kΩ resistor that can be connected in series to make it a 10 kΩ resistor, or in parallel to make it a 100 Ω resistor.

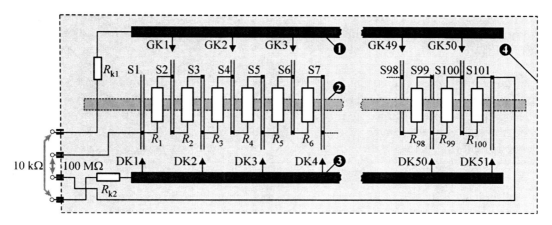

Figure 3.13 The VOO-100 (10 kΩ - 100 MΩ) Hamon resistor designed at the Primary Electromagnetic Laboratory of Croatia (GKx – upper contacts, DKx – lower contacts, Sx – connections)

3.1.7 Quantum Hall Resistance Standard

Under international agreement beginning January 1, 1990, the SI definition of the unit Ohm is related to the quantum Hall effect via the von Klitzing constant. This allows national laboratories to have their own quantum Hall resistance standard, which has a relative uncertainty of about 0.2 ppm to the definition of the SI unit Ohm. Today, national metrology institutes of the most developed nations have this standard. In 1988, the International Commission for Weights and Measures (CIPM) proposed that national laboratories use the same values for the von Klitzing constant. The suggested value is:

$$R_{k-90} = 25812,570 \; \Omega \quad [3.19]$$

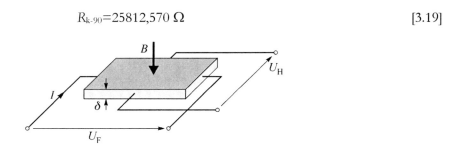

Figure 3.14 Quantum Hall effect

Von Klitzing constant is a universal quantity, associated with natural constants h and e, so that:

$$R_k = \frac{h}{e^2} \quad [3.20]$$

where e is the elementary charge and h is Planck's constant. Although the R_{k-90} quantity is associated with the von Klitzing constant, this does not mean that it is equal to $R_k = \frac{h}{e^2}$. Various laboratories in the world are performing experiments to determine the value of basic constants in nature. The quantum Hall effect was discovered by **von Klitzing** in 1980, for which he received the Nobel Prize for Physics in 1985. The heart of the quantized Hall resistance standard is the Hall plate, a planar MOSFET (Figure 3.14). It is designed to pass currents through a thin, semi-conductive layer.

A planar transistor is used at cryogenic temperatures from 1 to 2 K. A magnetic field B of a few tesla is applied perpendicularly to the Hall plate. Current is directed along the plate, and a Hall voltage U_h is measured perpendicularly both to the magnetic field and current. Because of the quantized action of electrons, the Hall voltage changes in steps of fixed values, rather than continually with the change of the magnetic field (Fig. 3.15), so that small changes in magnetic fields have no effect on the amplitude of the Hall voltage.

If the Hall voltage is U_h, R_{k-90} is a value associated to the von Klitzing constant, I_h is the current flowing through the MOSFET, i is the integer that denotes the Hall's step, where R_h is measured, and R_h represents a quantum Hall resistance, then the Hall voltage U_h and Hall resistance R_h ratio applies:

$$U_h = \frac{R_{k-90}}{i} \cdot I_h \quad [3.21]$$

$$R_h = \frac{U_h}{I_h} = \frac{R_{k-90}}{i} \qquad [3.22]$$

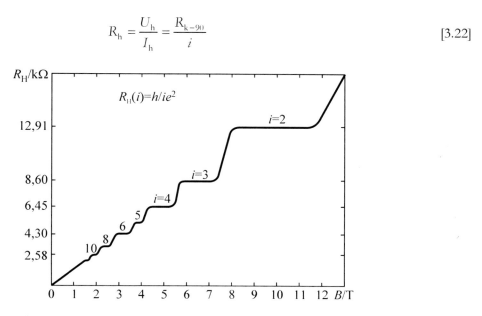

Figure 3.15 Hall resistance in relation to the magnetic field

The repeatability of results is twice better than the best methods for determining the absolute ohm. Specially developed methods are used to compare various other resistance standards in relation to one of the values of the quantized Hall resistance.

3.2 CAPACITORS

Capacitors are passive electronic components that have a pair of conductors separated by a dielectric (insulator). Voltage applied to the capacitor creates a static electric field in the dielectric that stores energy. The most important characteristic of the capacitor is its capacitance, measured in farads, a ratio of the electric charge on each conductor and the applied voltage. As capacitors pass only alternate current and block direct current, they are used in many different applications, such as filtering (to smooth the output of power supplies) and resonant circuits (radios). Capacitors also have some other important characteristics, and as resistors, are produced in a wide range of different types. Their other important characteristic is breakdown voltage. Also, dielectrics usually pass a small leakage current, and conductors also have some resistance.

3.2.1 Equivalent Circuit of Capacitors

Capacitors must meet the following requirements:

- capacitance well known
- time stable
- independent of frequency, voltage, and temperature
- as "clean" capacitance as possible
- high insulation resistance between electrodes
- low dielectric losses
- low self-inductance.

INSTRUMENTATION AND MEASUREMENT IN ELECTRICAL ENGINEERING

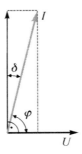

Figure 3.16 Capacitor losses (angle δ)

In an ideal capacitor, the phase shift φ between current and voltage should be 90°; however, imperfect capacitors have, due to the **capacitor loss angle δ**, (Figure 3.16), a phase shift of **90°-δ**. The losses in the capacitors are then:

$$P = UI\cos\varphi = UI\sin\delta \approx U^2 C\omega \sin\delta \approx U^2 C\omega\delta \qquad [3.23]$$

Such imperfect capacitors can be replaced by a serial combination of resistor R_S and capacitor C_S, or by a parallel combination of resistor R_P and capacitor C_P (Figures 3.17a and 3.17b). In practice, when the loss angle δ is small, the following applies:

$$C_S = C_P\left(1 + tg^2\delta\right) \approx C_P. \qquad [3.24]$$

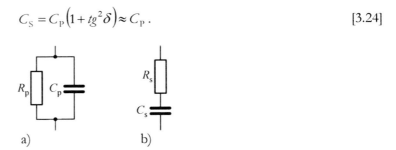

Figure 3.17 Equivalent circuits of the capacitor

Loss factor **tan δ** is usually marked with a **D** and is often determined by the following equation:

$$tg\delta = \frac{1}{R_P C_P \omega} = R_S C_S \omega \qquad [3.25]$$

3.2.2 Capacitor Standards

The calculable capacitor standard of Thompson-Lampard has already been mentioned. Such a standard is inappropriate for everyday use, so laboratories use the stand-alone standards of capacitance (Figure 3.18), which have excellent time stability, low loss angle, low temperature coefficient, and frequency independence. Such standards are calibrated by calculable standards in national metrology institutions to establish traceability. The best-known capacitors are **plate capacitors** with air or quartz as the dielectric. Standards with quartz as the dielectric usually have values of 10 pF and 100 pF. Often these standards are placed in thermostats to establish temperature stability. When mica is used as a dielectric, it is possible to achieve greater values of capacitance. In **capacitance decade boxes**, different values are obtained using parallel connection of individual capacitors. In high voltage measurement applications, special types of capacitors are used that have a small loss angle δ. Such capacitors are filled with compressed gas (e.g. nitrogen) under pressure of 10-15 atmospheres. Recently, sulfur hexafluoride (SF_6) has been used, which needs several times lower pressure than nitrogen for the same breakdown voltage.

Figure 3.18 Usable standard capacitor 100 pF (Primary Electromagnetic Laboratory of Croatia)

Capacitance standards are most often found in metal casings; these standards have three connections (the third terminal is connected to the grounded housing (Figure 3.19). In addition to the capacitance between electrodes (C_{12}), there are also parasitic capacitances of C_{10} and C_{20} between the electrodes and casing. These capacitances can reach several tens of picofarads, and must be taken into consideration when performing precise measurements.

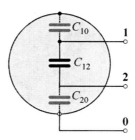

Figure 3.19 Parasitic capacitances of capacitor

There are three possible ways of connection. In the first case, terminal 1 is attached to the earth potential, so the total capacitance of the standard is then $C = C_{12} + C_{10}$. In the second case, the terminal 2 is connected to the earth potential and the total capacitance is then $C = C_{12} + C_{20}$. If no terminal is connected to the earth potential, then the total capacitance will be equal to $C = C_{12} + C_{10} \cdot C_{20}/(C_{10}+C_{20})$.

3.3 INDUCTORS

Inductors are passive components usually made from conducting wire shaped as a coil. Inductors can also store energy just like capacitors, but they work differently. The energy stored in inductors comes from the magnetic field created by the passing of the electric current through the coil wire. The most important characteristic is its inductance, which is measured in henries. Inductors are also common components in electric circuits, with different applications.

3.3.1 Equivalent Circuit of Inductors

Similar to capacitors, inductors must satisfy the following characteristics:

- their inductance should be exactly known and constant
- they should be independent of frequency, temperature, and external magnetic fields
- the resistance R of the inductor is as small as possible, and that the inductor time constant (L/R) is as high as possible
- self-capacitance is as low as possible.

Inductance standards are always wound on a non-ferromagnetic body in order to avoid the influence of the nonlinear dependence of permeability on the current that flows through the coil and the losses due to hysteresis and eddy currents. Therefore, the required amount of inductance is achieved only with a lot of turns, which increases the effective resistance R of the coil, and thus reduces the **time constant** of coil L/R.. For alternating current, the inductor is characterized by a **quality factor** $Q=L\varpi/R$. It is therefore necessary to know the resistance R of the coil and take it into account during measurement. It is also important that the effective resistance is independent of frequency, and the coils are wound with thin isolated wires.

The equivalent circuit of the inductor is similar to the resistor, as resistance and inductance are connected in serial, and self-capacitance is connected in parallel (Fig. 3.20).

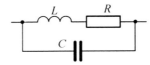

Figure 3.20 Equivalent circuit of the inductor

3.3.2 Inductance Standards

Calculable self-inductance standards are not suitable for daily measurements due to large dimensions and weight, and low inductance.

Figure 3.21 Inductance standard of 0.01 H

Their inductance is determined by the number of turns, and by applying very precise dimension measurements. They are used only for calibration of usable inductance standards. **Usable inductance standards** (Figure 3.21) are smaller, usually multilayer coils wound round the body (ceramic, marble, quartz, etc.) They are manufactured in decimal grading from 0.1 mH to 1 H. **Standards of mutual inductance** have practically the same shape as inductance standards, but the primary and secondary coils are wound together, so that they lie next to each other. Therefore, mutual inductance is equal to self-inductance of the primary and secondary coil.

3.4 LABORATORY VOLTAGE SOURCES

Laboratory voltage sources can be divided into two groups:

- AC
- DC

DC sources can be accumulators, batteries, rectifiers, and generators. Batteries and accumulators can be either primary, which cannot be recharged after discharge, or secondary, which can be recharged after discharge. **Batteries** have a constant voltage, very low internal resistance, and are suitable for precise measurements. Some of the best known are lead, nickel-cadmium, and silver-zinc batteries. **Rectifiers** are practical because the DC voltage and current are received from the power electricity grid. The rectified voltage is smoothed by filters. **Generators** (DC machines) are used in measurements when larger powers are needed.

For the AC source, the power grid is always available, but with restrictions, since the voltage can be changed up to 10% from the nominal value of 230 V, which is sometimes unsuitable for precise measurement purposes. AC generators are used for higher power purposes, just like DC generators. **Electronic sources** are also available for different frequencies, voltages, and wave shapes.

Sine wave generators generate sinusoidal voltage waveforms of variable frequency and amplitude. **Function generators** usually generate rectangular, triangular, and sine waveforms. **Signal generators** allow amplitude and frequency modulated signals. **Calibrators** are generators of alternating voltage with small error limits, which allow fine tuning of the amplitude and frequency (for AC) of their output signals. They are also used for the calibration of instruments.

3.5 VOLTAGE STANDARDS

3.5.1 Josephson Array Voltage Standard

The Josephson Voltage Standard is one of the quantum standards. It is based on the AC Josephson effect, which produces voltage directly proportional to frequency, a quantity which can be very accurately measured. The connection between two conductors in the superconducting state, separated by a thin layer of insulation (order of magnitude of nanometers), is radiated with the high frequency field of frequency f (order of magnitude - gigahertz). Then, the DC voltage step is dependent of the current passing through the junction. The difference between the two steps is then:

$$U_0 = \frac{h}{2e} f = \frac{f}{K_J}, \qquad [3.26]$$

where h is the Planck constant, e is the charge of the electron, and $K_J = \dfrac{2e}{h}$ is the Josephson constant (since January 1, 1990, the recommended value of the Josephson constant is $K_{J\text{-}90} = 483\,597{,}9 \cdot 10^9$ Hz/V). The current through the circuit corresponding to n-th step will generate a voltage:

$$U_n = nU_0 = \frac{nf}{K_J}. \qquad [3.27]$$

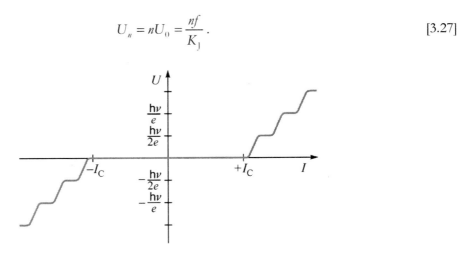

Figure 3.22 Josephson Voltage Array Standard (JAVS) DC voltage steps in relation to the applied current passing through junction

The voltage per junction is about 500 µV, so higher voltages are achieved by cascading Josephson junctions. This voltage does not depend on the type of semiconductor, nor on the temperature inside the superconducting region, nor on the method of isolating superconductors. Various international laboratories reproduced voltages with uncertainty of only $1 \cdot 10^{-8}$. Commercial Josephson voltage standards give a maximum voltage of 10 V. These devices, just like the Quantum Hall Resistance Standard, have a high purchase and maintenance price, especially for keeping the Josephson junctions on temperatures close to absolute zero (273.16 °C), while operating the JAVS. Therefore, only several national metrology institutes in the world possess the Josephson voltage standard. Even at these institutes, it is not used year-round, but only occasionally for assigning voltages to cheaper electronic voltage standards.

3.5.2 Weston Voltage Standard

Weston cells are the oldest voltage standards in the world. The Weston Voltage Standard was accepted as the international standard voltage in 1908. Sets of Weston cells were used for decades in the national metrology laboratories as DC voltage standards.

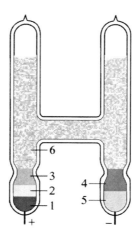

Figure 3.23 Weston cell

The Weston cell is a chemical source made of a glass container in the shape of the letter H (Figure 3.23). The positive electrode consists of mercury (1), on top of the mercury is the depolarization paste made of mercury sulfate (2), and the negative electrode is made of cadmium amalgam (5). On top of the electrodes are crystals of cadmium sulphate (3 and 4) and cadmium sulphate solution (6).

The electrolyte is a solution of cadmium sulphate in distilled water. In the careful preparation of this cell, the voltage is **1.01865 V**, at **20 °C**. The cell voltage is dependent on temperature, and the voltage in the temperature range of 10°C to 25 °C is determined from tables. The cell voltage is also dependent on the load. The loading and temperature hysteresis (first increasing and then decreasing the load or temperature) leads to changes in the voltage cell, so there is a need to wait for some time (from several minutes to several months, depending on the duration of load or temperature changes) before the cell reaches the initial value of voltage. Therefore, Weston cells should be used as voltage references in compensation circuits. The Weston cell is never used as one standard, but several standards are usually placed in a thermostated container, such as presented in Figure 3.24.

Figure 3.24 Weston cell standard at the Croatian Primary Electromagnetic Laboratory

3.5.3 Electronic Voltage Standards

Electronic voltage sources have the **Zener diode**, which functions as an ordinary diode in forward direction, but also permits current in the reverse direction if the voltage is larger than the breakdown voltage known as Zener knee voltage U_Z. The diode current in the reverse direction rises slowly, and when the voltage reaches Zener voltage, it rises rapidly. The Zener voltage depends on the diode construction, but it is usually several volts (Figure 3.25). The main characteristic of Zener voltage, which is essential for the production of voltage standards, is its time stability. To use the Zener diode for the construction of the voltage standard, its slope characteristics or dynamic resistance R_Z must be in the area of interest:

$$R_Z = \frac{\Delta U_Z}{\Delta I_Z} \qquad [3.28]$$

which ranges from 0.5 to 150 Ω. The principal circuit of voltage standard with the Zener diode is shown in Figure 3.26. It is evident that this circuit stabilizes the voltage that is applied to the input. If stabilized output voltage is needed, which serves as the voltage standard, it is necessary that the relative change of input voltage corresponds to the less relative change of output voltage. The ratio of relative change of input and the relative change of output voltage is called the **factor of stabilization**:

$$S = \frac{\Delta U_{ul}}{U_{ul}} : \frac{\Delta U_{iz}}{U_{iz}} = \frac{U_{iz}}{U_{ul}} \cdot \frac{\Delta U_{ul}}{\Delta U_{iz}} = \frac{U_{iz}}{U_{ul}} \cdot \left(1 + \frac{U_{ul} - U_{iz}}{R_Z(I_{IZ} + I_Z)}\right). \qquad [3.29]$$

The factor of stabilization can be as high as 100, which means that a change of input voltage by 10% causes a change in output voltage of 0.1%. If such stabilization is too small, it can be increased by cascading multiple steps. Then the total stabilization factor is:

$$S_{uk} = S_1 \cdot S_2 \cdot ... \cdot S_n. \qquad [3.30]$$

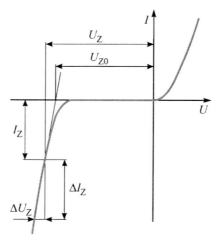

Figure 3.25 U-I characteristic of Zener diode

MEASURING ELEMENTS

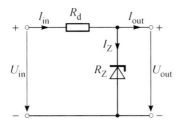

Figure 3.26 Principal circuit of Zener voltage standard

However, increasing the number of steps also increases the required input voltage. The Zener voltage standard temperature coefficient is smaller than the temperature coefficient of the Weston cell, and it can be further reduced by temperature compensation by choosing the diodes with a small temperature coefficient. It is also possible to use thermostated casings to reduce temperature effects. Figure 3.27 shows the Zener voltage standard at the Croatian Primary Electromagnetic Laboratory.

Figure 3.27 Zener voltage standard at the Croatian Primary Electromagnetic Laboratory

3.6 ADJUSTING THE CURRENT

Sources generally have the option of tuning the output voltage. It is often necessary to change the current as well. Current adjustment is done using slide resistors, resistive decades, and in AC circuits, **variable transformers** are also used. In most cases, it is desirable to achieve a linear change of current. Slide resistors or decade boxes can be connected in two ways.

3.6.1 Adjusting the Current with Potentiometer

Figure 3.28 shows how to adjust the current with the potentiometer resistor R_{12}. The current is regulated by moving slider 3 of the resistor R_{12}. Load R is connected in parallel to the resistance R_{13}. With $R_{12} < R$, a nearly linear change of current is achieved by increasing the resistance R_{13}, i.e., moving the slider 3 of the resistor R_{12}:

$$I_T \approx \frac{U}{R} \cdot \frac{R_{13}}{R_{12}}. \qquad [3.31]$$

This circuit is used to adjust the current of low power loads, because the ratio between power on load and power provided by the source is satisfactory.

INSTRUMENTATION AND MEASUREMENT IN ELECTRICAL ENGINEERING

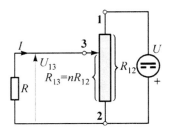

Figure 3.28 Adjusting the current with the potentiometer

3.6.2 Adjusting the Current with Resistor

Figure 3.29 shows another method of current adjustment, where resistance R_{13} is connected in series with the load R. With $R_{12} = R$, the current adjustment is approximately linear for currents in the range of $I_T = \dfrac{U}{2R}$ to $I_T = \dfrac{U}{R}$. This circuit is used for high power loads.

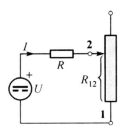

Figure 3.29 Adjusting the current with the potentiometer in series connection

Selected bibliography:

1. A. F. Dunn: *"Primary Electrical Units at the National Research Council of Canada"*, Metrologia, March 1968, pp. 180-184.
2. V. Bego, *"Measurements in Electrical Engineering"*, (in Croatian) Tehnička knjiga, Zagreb, 1990
3. Fluke Electronics, *"Calibration: Philosophy in Practice"*, Second Edition, 1994.
4. M. E. Cage, D. Yu, B. M. Jeckelmann, R. L. Steiner, R. V. Duncan: *"Investigating the Use of Multimeters to Measure Quantized Hall Resistance Standards"*, IEEE Trans. Instrum. Meas. IM-40 2, pp. 262-266., 1991.
5. D. Braudaway: *"Precision Resistors: A Review of Material Characteristics, Resistor Design, and Construction Practices"*, IEEE Trans. Instrum. Meas. Vol. 48, No. 5, pp. 878-883, October 1999.
6. Witt TJ.: *"Electrical Resistance Standards and the Quantum Hall Effect"*, Review of Scientific Instruments, Vol. 69, No. 8, pp. 2823-2843, August 1998.
7. National Physics Laboratory, *"Electromagnetic Metrology, DC & Low Frequency Measurement Services"*, 1999.
8. C. H. Page: *"Errors in the Series-Parallel Buildup of Four-Terminal Resistors"*, Journal of Research of the NIST, Vol. 69C, No. 3, pp. 181-189, July-September 1965.
9. Hamon B. V., *"A 1-100 Ω Build-up Resistor for the Calibration of Standard Resistors"*, Jour. Sci. Instr., vol. 31, pp. 450-453, 1954
10. I. Peterson: *"Physics Nobel Spotlights Quantum Effect"*, Science News, October 1998
11. R. F. Dziuba: *"Resistors Encyclopedia of Applied Physics"*, VCH Publishers Inc., New York, pp. 423-435, July 1996.

12 C. M. Sutton, M.T. Clarkson: "*A General Approach to Comparisons in the Presence of Drift*", Metrologia, Volume 30, Number 5, January 1994, pp. 487-494.
13 A. F. Dunn: "*Methods of Intercomparing a Group of Standards*", APEM-797, National Research Council of Canada, March 1961.
14 D. Vujević: "*The Contributions to the Maintenance of Voltage Standard*", doctoral dissertation, University of Zagreb 1982.
15 R. Malarić, "*Maintaince of Croatian Resistance Standads*", doctoral dissertation, University of Zagreb 2001.
16 J. Butorac and I. Kunšt: "*Achievement of 0,1 ppm accuracy with the resistance standard of 100 MΩ*" (in Croatian), Proc. of the JUKEM '90, Part 1, pp. 127−133, Sarajevo, 1990.
17 I. Leniček, R. Malarić, A. Šala, "*Calibration of 100 MΩ Hamon resistor using current-sensing Wheatstone bridge*", Proc. of the 13th IMEKO TC4 Symposium, pp. 75−79, Athens, 2004.
18 *Hamon Resistor Transfer Standard*, Features, Guildline, http://www.guildline.com/9350.html, Retrieved October 10th, 2010

4. ANALOGUE MEASURING INSTRUMENTS

For measuring electrical quantities, measuring instruments are needed, since human senses are not appropriate for that purpose. The basic operation of these instruments depends on the type of measurement system used. Different analogue instruments base their principle of operation on various physical laws and phenomena. Some of the instruments are appropriate for measuring alternate currents, and some are better for measuring direct currents. However, most analogue instruments have some common features, mainly their conversion of measured values to the mechanical force that causes the deflection of the instrument pointing device from which the measured value can be read. Even though analogue instruments are sometimes replaced by digital instruments, they are still used in many applications today, such as control, distribution, and automation panels (remember also your car speedometer). The operator can monitor many process quantities on analogue instruments more quickly and with greater visual ease than when using digital instruments. The operator can thus interpret data to detect the installation status, as the position of the needle is often the indication of malfunction. Analogue instruments are also manufactured to operate in harsh and demanding environments. A final important advantage is that they usually do not need auxiliary power supplies.

4.1 BASIC CHARACTERISTICS OF ANALOGUE INSTRUMENTS

4.1.1 Torque and Counter-Torque

To achieve a constant deflection of the instrument pointing device, two torques are necessary: M_1 **torque** produced by the measured electrical quantity X (voltage, current, etc.) and **counter-torque** M_2, which opposes torque M_1. The counter-torque is obtained by a coil spring and depends only on the deflection angle of the pointing device. Torque M_1 is usually a function of measurand X and a deflection angle α, while counter-torque M_2 depends only on deflection angle α:

$$M_1 = f(X, \alpha) \qquad [4.1]$$

$$M_2 = f(\alpha). \qquad [4.2]$$

By equaling torques ($M_1 = -M_2$), the deflection system will have a constant deflection, and this deflection will be a function of the measured quantity X:

$$\alpha = f(X). \qquad [4.3]$$

The coil spring that creates the counter-torque also serves to return the pointing device back to the starting position when the instrument is disconnected from the measured quantity. When starting the measurement, the deflection system must move to the final (measurement) position as quickly as possible. Therefore, the oscillations of the deflection system must be reduced by damping. This can be done electromagnetically, by air or by liquid. The resulting torque is called the **damping torque**. To analyze the motion of the deflection system, the **degree of attenuation s** is used. Depending on the degree of damping, three types of motion of the deflection system can be distinguished (Fig. 4.1.):

- **non-attenuated vibrating** ($s=0$)
- **attenuated vibrating** ($0<s<1$)
- **aperiodical** ($s>1$)

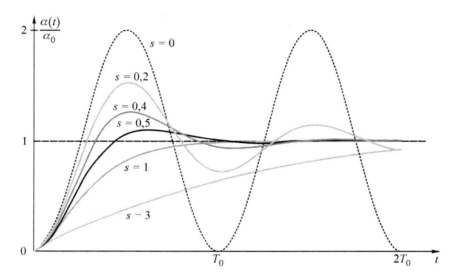

Figure 4.1 Types of motion of the pointing device

Non-attenuated vibrating motion is only theoretically possible because the system is moving in the air, so there will always be some damping torque. In such a hypothetical situation, the deflection system would oscillate from zero to the double value of the measured quantity, without ever settling to the desired **final deflection.** For measuring instruments with the degree of attenuation $s>1$ (aperiodical), the deflection system would take infinitely long to make it to the final deflection. It is therefore necessary that the instrument should have an attenuated vibration motion. Usually the degree of attenuation s is chosen in such a way that the amplitudes of vibration reach values that correspond to the accuracy class of the instrument as quickly as possible.

4.1.2 The Scale and Pointing Device

All electrical measuring instruments have a scale which has division lines and associated numbering. We distinguish between the **indication** and **measuring** range of the measuring instrument. The indication range covers the entire length of the scale at which measurements can be made, while a measuring range is the range in which the measuring instrument meets the required accuracy and is generally smaller than the indication range (Figure 4.2.b).

ANALOGUE MEASURING INSTRUMENTS

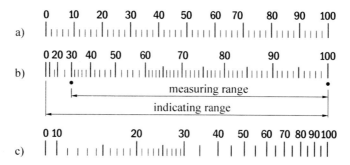

Figure 4.2 Typical scales of analogue instruments

The scale has **linear** characteristics (Figure 4.2.a) when the torque has linear dependence on the measured quantity, or **square** characteristics (Figure 4.2b) when the torque depends on the square of the measured quantity. Also, the scale can have other different characteristics, e.g. **logarithmic** (Figure 4.2.c).

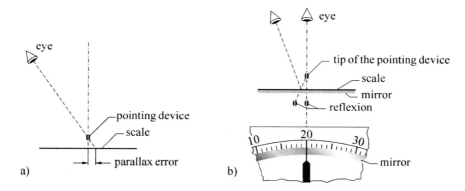

Figure 4.3 Parallax error

Pointing devices in the measuring instruments are usually hard, although light pointing devices can be found in older instruments. The pointing device is usually distanced from the scale, so there is the possibility of error due to parallax if the measurer does not look perpendicular to the scale (Figure 4.3a). To avoid this effect, scales with mirrors can be used (Figure 4.3b).

4.1.3 Uncertainty in Reading Analogue Instruments

With precise analogue measuring instruments, the uncertainty of reading **n** is about one-tenth of the minimum scale spacing. In absolute terms, the limits of error are the same along the entire scale, but if expressed as a percentage of the measured value, they become larger as the deflection of the pointer becomes smaller. Percentage uncertainty of readings ε_n is inversely proportional to the deflection α of the pointing device:

$$\varepsilon_n = \frac{n}{\alpha} \cdot 100\% \qquad [4.4]$$

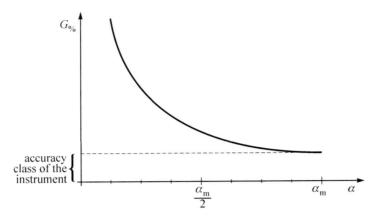

Figure 4.4 The reading error of analogue instruments is reduced as the measured value is closer to the maximum deflection of a given range

The smallest percentage uncertainty of readings is achieved with the maximum deviation α_m, so the measuring range of the instrument must be selected in such a way that the reading is close to the maximum deflection (Fig. 4.4).

4.1.4 Sensitivity and Analogue Instrument Constant

The sensitivity of the analogue instrument is defined as the ratio of pointing device deflection change $\Delta\alpha$ and change of the measurand ΔX:

$$S = \frac{\Delta\alpha}{\Delta X}.$$ [4.5]

In the linear scale, the sensitivity is equal throughout the length of the scale, while in the quadratic scale, the sensitivity of the scale increases from the start to the end. The sensitivity of measuring instruments depends on the power necessary for their deflection. The consumption depends on the current passing through the instrument and its voltage drop. Currents through the instrument that have high internal resistance (e.g., voltmeter) will be smaller than the currents through the instrument that have small internal resistance, and such an instrument will have a large current sensitivity. In analogy to this, a small internal resistance instruments (e.g. ammeter) will have significant voltage sensitivity.

The instrument constant is the reciprocal of sensitivity, and it is calculated by dividing the measuring range expressed in units of measurement by the total number of scale division lines.

4.1.5 Standards and Regulations for the Use of Analogue Electrical Measuring Instruments

The use, procurement, and manufacture of instruments is greatly eased by defining basic terms, tags, testing methods, use, and limit errors. All countries adopt their own standards, as recommended by the International Electrotechnical Commission (IEC). According to the current standards, analogue measuring instruments are classified into eight classes of accuracy (Table 4.1). The accuracy class is equal to the percentage error of the measurement range. The instrument used in the measuring range and in referent conditions must not exceed the limits of errors listed in Table 4.1. in any part of the measurement range. For example, the error of voltmeter with accuracy class 1, on the 10 V measuring range must not exceed the value of ± 0,1 V (1% of 10 V). This means that the percentage error at full deflection will be smallest (10 V ±0,1 V or 10 V ± 1 %), while, for example, in the middle of a scale, the instrument will have twice the percentage error (5 V ± 0,1 V ili 5 V ± 2 %).

Table 4.1. Classification of measuring instruments – accuracy class

Limits of error of measuring instruments, depending on the accuracy class							
Accuracy class	0,05	0,1	0,2	0,5	1	1,5	2,5	5
Limits of error from the range	±0,05%	±0,1%	±0,2%	±0,5%	±1%	±1,5%	±2,5%	±5%

In addition to the measuring quantity, other physical quantities can influence the deflection of the instrument; these quantities are called **influential factors**. These may include ambient temperature, humidity, position of the instrument, exterior magnetic and electric fields, shape of measured current, and the power factor. Influential factors affect the accuracy of the instruments; therefore, standards prescribe the allowed values of certain influential quantities. These values are called **reference values**, and are expressed with a certain tolerance. Measuring instruments will operate according to the accuracy class only if the reference values of influential quantities are met. For example, the effects of temperature on the Class depend on the scope of the measurement. In general, instruments maintain their Class between +10 and +30 °C. This interval can be smaller for precise applications. In these cases, the limit values are indicated on the scale.

The standards also prescribe the data and information each measuring instrument must have on the scale or on the outside of the housing. The most important data are:

1. manufacturer's name or logo
2. measured unit, denoted by symbol (e.g., A, V, W, Hz)
3. serial number
4. accuracy class (Table 4.1.)
5. current type (Table 4.2.)

Table 4.2. Examples of symbols for some types of current

DC	—
AC 1-phase	∼
DC and AC	≂
AC 3-phase	≋
Non-symmetrical AC 3-phase	≋

6. test voltage (which examines the isolation of the measuring instrument between the casing and measuring system; symbol ☆ indicates a test voltage of 500 V, while ☆3 indicates the test voltage of 3 kV.)

7. type of measuring system (the symbol for the moving coil instrument is ⌂, an instrument with a moving iron has the symbol ⚏, while the symbol for the electro-dynamic instrument is ⏚.)

4.2. INSTRUMENT WITH MOVING COIL AND PERMANENT MAGNET (IMCPM)

Due to extremely good features, this instrument is used frequently, mostly in universal instruments (Figure 4.5). The coil is wound onto the moving aluminum frame, which is moving in the air gap between the poles of permanent magnets. The air gap has a practically homogeneous magnetic field in the direction perpendicular to the air gap. When current flows through the coil, the conductors in the coil that are in the air gap feel the force that acts in tangential direction towards the conductors. This force creates a torque proportional to current:

$$M_1 = B \cdot n \cdot b \cdot h \cdot i = G \cdot i. \qquad [4.6]$$

Counter to this torque acts the counter-torque, which is proportional to the deflection angle of the moving coil:

$$M_2 = D \cdot \alpha. \qquad [4.7]$$

With constant current, the moving part of the instrument will make it to the final position where both torques are equal:

$$M_1 = -M_2. \qquad [4.8]$$

Then the deflection is proportional to the **mean value** of the current:

$$I = C_i \cdot \alpha. \qquad [4.9]$$

where C_i is the **current constant** of instrument.

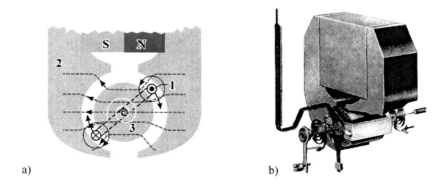

Figure 4.5 Instrument with moving coil and permanent magnet (IMCPM)

The scale of this instrument is **linear**. Stationary deflection will be obtained only for the DC current. In AC currents, the torque M_1 changes direction in every semi-period, so the moving part of the instrument (pointing device), depending on the frequency, will more or less oscillate around the zero position. At higher frequencies, this oscillation can no longer be observed due to inertia of the eye. Two spiral springs create the antagonistic pair to move the needle to the position zero.

ANALOGUE MEASURING INSTRUMENTS

4.2.1 Extending the Measurement Range

To achieve full deflection of a basic analogue instrument, it takes less than **20 mA** in most cases. To measure larger currents, it is necessary to extend the measurement range of the instrument. Since IMCPM is also used as a voltmeter (voltage measurement using Ohm's law is reduced again to the measurement of currents), there is a difference between extending a current and voltage ranges.

Extending the voltage measurement range can be achieved by connecting the resistor R_P in series with the coil of the measuring instrument (Figure 4.6). Resistor value R_P can be calculated using the formula:

$$R_P = \frac{R_V}{U_V}(U - U_V) \qquad [4.10]$$

where R_V is the internal resistance of the instrument, U_V is the voltage drop on the instrument, and U is the voltage that we want to measure. The quotient R_V/U_V is called characteristic resistance of the instrument and is expressed in Ω/V; it is the resistance needed to increase the measurement range of the voltmeter by 1 V. When a voltmeter has multiple measurement ranges, there must be more resistances that are connected in series and selected by switches.

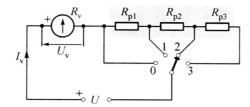

Figure 4.6 Extension of the voltage range

Extending the current measurement range is achieved by connecting resistors, called shunts, parallel to the instrument so that only a part of the measured current passes through the instrument. Shunt resistance R_S can be calculated according to the following formula:

$$R_S = \frac{R_V I_V}{I - I_V} \qquad [4.11]$$

where R_V is the internal resistance of the instrument, I_V is the current that passes through the instrument, and I is the measured current. To extend current measuring range, more shunts connected in parallel can be used (Figure 4.7).

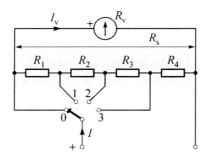

Figure 4.7 Extension of the current range

4.2.2 Measurement of Alternating Current and Voltage Using IMCPM

Due to small consumption and high sensitivity, IMCPM is often used for the measurement of alternating current and voltage.

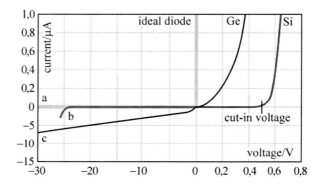

Figure 4.8 Static characteristics of semiconductor rectifiers

For this purpose, it is necessary to rectify the AC current using rectifiers. In the ideal case, rectifiers pass the current in only one direction, while in the other direction they exhibit infinite resistance.

The **mechanical** rectifier is closest to the ideal rectifier (curve **a** in Figure 4.8). There are cheaper versions using **semiconductor** rectifiers (silicon or germanium diode - curve **b** and **c** in Figure 4.8). To a certain voltage in the forward direction, which is called the **cut-in voltage**, the current through the diode practically equals zero (for the silicon diode the cut-in voltage is approximately 0.6 V).

Rectifying can be half-wave, full wave, and peak.

a) Half-Wave Rectifier

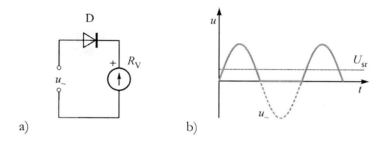

Figure 4.9 Half-wave rectifier

If the diode is connected in series with the instrument (Figure 4.9.a), and if the voltage source is connected parallel to it, the instrument will measure the **average value** of the rectified voltage:

$$U_{AVG} = \frac{U_M}{\pi}. \qquad [4.12]$$

During the positive half-wave period, the current flows through the diode **D**, while during the negative half-wave period, the current through the instrument is practically zero. In Figure 4.9.b the sinusoidal input current waveform and the rectified current is shown.

b) Full-Wave Rectifier

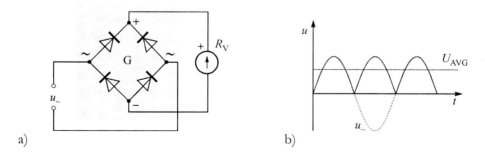

Figure 4.10 Full-wave rectifier

To obtain greater sensitivity of the instrument, the **Graetz** circuit can be used to achieve a full-wave rectifier circuit (Figure 4.10.a). Then, the rectified sinusoidal signals waveform looks like Figure 4.10.b, and the mean voltage value is:

$$U_{AVG} = \frac{2U_M}{\pi}. \qquad [4.13]$$

Since in most cases the **effective** value, and not the mean value, is of interest when measuring the AC currents and voltages, such instruments are sometimes calibrated in effective values of **sinusoidal** waveforms. The effective value of voltage or current is the value that produces the same amount of heat as the DC voltage would produce in the same time period. The waveform signal measured is usually sinusoidal; however, if the measured signal is of other shapes, an error will arise depending on the **shape factor** ξ of the measured signal. The shape factor is the ratio of the effective and mean value of the measured quantity. The shape factor of the sinusoid is ξ_0=**1.11**, and the instrument that is calibrated in the effective value of sinusoids, when measuring voltage, will show $U_{AVG}*\xi_0$ and not correct $U_{AVG}*\xi$. Therefore, the percentage error can be calculated:

$$p = \frac{\xi_0 - \xi}{\xi} \cdot 100\%. \qquad [4.14]$$

For example, if the triangular waveform has a shape factor ξ=1,15, by measuring the voltage using a rectified **IMCPM** instrument calibrated in the effective value of sinusoids, there will be a percentage error $p = \frac{1,11 - 1,15}{1,15} \cdot 100\% = -3,96\%$.

Table 4.3 shows the most frequent waveforms together with the shape factor ξ and crest factor σ, which is calculated as the ratio of the peak to RMS value.

Table 4.3 Waveforms, shape factor ξ and crest factor σ

Waveform		ξ	σ	Waveform		ξ	σ
Sine		1,11	1,41	Sawtooth voltage		1,155	1,73
Half-wave rectified sine		1,57	2	Triangular voltage		1,155	1,73
Full wave rectified sine		1,11	1,41	Rectified rectangular voltage		1,41	1,41
Rectangular voltage		1,0	1,0	Rectified triangular and sawtooth voltage		1,15	1,73

c) Peak Rectifier

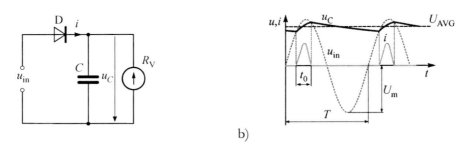

Figure 4.11 Peak rectifier

If a capacitor is added in parallel to the above IMCPM instrument with rectifier (Figure 4.11.a), it will charge approximately to the **peak** value when a current passes through the diode, and will then discharge through the measuring instrument. Figure 4.11.b shows the waveform of voltage and current with a peak rectified instrument, where u_C shows the voltage across the capacitor C, and U_{AVG} its mean value, practically equal to the peak value of measured voltage u_{in}. Therefore, this instrument will have a deflection proportional to the peak value regardless of the shape factor. As the current-voltage characteristic of the diode is non-linear, the scale of the instrument with a rectifier will also be **non-linear**.

The scale can be linearized, at the expense of sensitivity, adding a sufficiently large resistance in series with a rectifier (Figure 4.12).

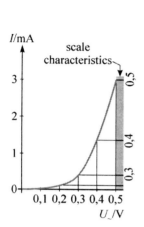
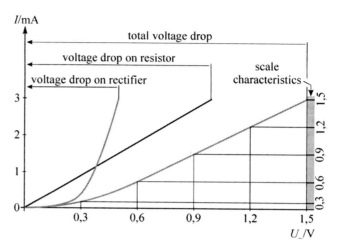

Figure 4.12 Linearization of scale

4.2.3 Universal Measuring Instruments
Universal instruments are instruments that usually have IMCPM as the instrument of choice, and can measure AC and DC currents and voltages in several ranges, and also resistances and some other quantities as well.

4.3 INSTRUMENT WITH MOVING IRON
Instruments with moving iron measure the **real effective value** and have a scale with square characteristics. The deflection is proportional to the square of the input voltage:

$$U_i = k \cdot u_u^2 \qquad [4.15.]$$

This means that the measurement of the DC voltage $\pm U$ or AC voltage of the same effective value will result in the same deflection on the instrument, so it is used primarily for the measurement of AC quantities. It is composed of a coil that transmits the measured current, with a fixed iron element and moving iron element in its centre, connected to the instrument's axis and needle. The indicator moves by the repulsion effect produced by the magnetic field between both iron elements, and the magnetic field depends on the current transmitted by the coil (Fig. 4.13). Thus, the measured current creates a torque in the instrument:

$$M_1 = \frac{1}{2} \cdot \frac{\Delta L}{\Delta \alpha} I^2. \qquad [4.16]$$

INSTRUMENTATION AND MEASUREMENT IN ELECTRICAL ENGINEERING

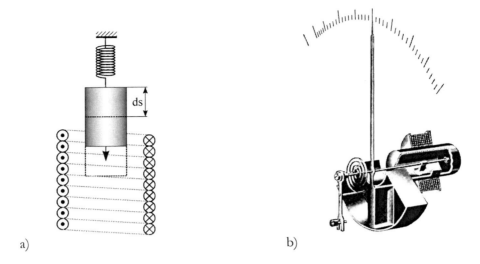

Figure 4.13 Instrument with moving iron

The torque is proportional to the square of the current through the coil and the change of its inductance due to the change of rotation of moving part $\frac{\Delta L}{\Delta \alpha}$. The counter-torque is produced by coil springs. If $\frac{\Delta L}{\Delta \alpha}$ is constant, then the scale will have square characteristics because the torque is proportional to the current squared. For the same reason, the instrument with a moving iron response is proportional to the **effective value** of alternating current, unlike **IMCPM** which has a response proportional to the **mean value**. Using the readings from these two instruments, it is possible to calculate the shape factor of a signal.

4.4 ELECTRODYNAMIC INSTRUMENT
Electro-dynamical instruments have a **moving coil** placed in the magnetic field of a **fixed coil**.

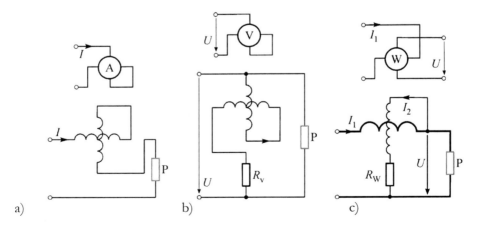

Figure 4.14 Different types of measurement instrument that can be achieved with the electrodynamic instrument

ANALOGUE MEASURING INSTRUMENTS

The fixed coil is wound with a small number of turns of thicker wire, and the moving coil is wound with a larger number of thinner wires. Coils can be connected in different ways, and so electro-dynamic instruments can be used to measure current, voltage, and power. Electro-dynamic instruments have a torque equal to the product of currents through the coils and their phase shift:

$$M_1 = k \cdot i_1 \cdot i_2 \cos\varphi \qquad [4.17]$$

Current measurements are performed according to Figure 4.14a. Both coils are connected in series, the current of the load flows through them, so the torque is proportional to the square of current. It follows that such an instrument will measure the effective value of current. For measuring voltage, a circuit from Fig 4.14b is used. Coils and resistance R_P are connected in series and then parallelly to the measured voltage U. Similar to the measurement of current, the torque is proportional to the square of voltage measured by the instrument, while the scale also has square characteristics.

In Figure 4.14c, the principle circuit of the electro-dynamic wattmeter is shown, where the current I flows through the **fixed coil**, while the **moving coil** is connected in series with resistance R_W and paralelly to voltage U. The torque is then proportional to the power of load P ($M_1 = k \cdot P$). While the voltmeter and ampermeters based on electro-dynamic instruments are rare today, electro-dynamic wattmeters are common, both in laboratories and in the industry.

4.5 ELECTRICITY METERING

Electric meters are instruments that are used to measure the consumption of electricity. They may be divided into DC (electrolytic, electro-dynamic, etc.) and AC (induction and electronic). AC meters are manufactured in large batches, so the manufacturer must strictly take care of their accuracy, since it directly influences the amount a customer pays to the supplier of electricity each month.

4.5.1 Induction Meter

In Figure 4.15, the induction meter is shown. It consists of two electromagnets: one of them - the voltage electromagnet - is wound with many turns of thin wire and connected to the power network voltage; the second, the current electromagnet, is wound with a small number of turns of thick wire through which the current of the load flows. Electromagnets are located opposite one another, with a round aluminum plate between them.

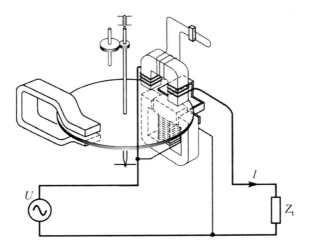

Figure 4.15 Induction electricity meter

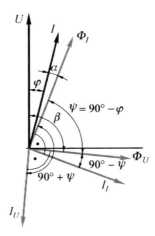

Figure 4.16 Vector diagram of the induction meter

The aluminum plate affects the magnetic fluxes in Φ_U and Φ_I of the electromagnet (Figure 4.16), which are proportional to the voltage and current of the load. These fluxes in the plate induce voltages that are lagging behind the fluxes by 90°. The voltages are inducing eddy currents I_U and I_I, and because of the small reactance of the plates, the eddy currents are almost in phase with the voltages. The flux Φ_U with current I_I creates one **torque,** and the flux Φ_I with current I_U creates another **torque**.

The total **torque** is then:

$$M_1 = k_1 \Phi_I \Phi_U \sin \psi \qquad [4.18]$$

where ψ is the phase shift between Φ_U and Φ_I. As the fluxes Φ_U and Φ_I are proportional to the voltage and current, we can write:

ANALOGUE MEASURING INSTRUMENTS

$$M_1 = k_1 UI \sin\psi \qquad [4.19]$$

To measure the active power, it is necessary that the torque be proportional to the consumed power $P = UI\cos\varphi$. This is achieved only if $\sin\psi = \cos\varphi$ or $\psi = 90° - \varphi$. Then the following will apply:

$$M_1 = k \cdot U \cdot I \cdot \sin\psi = k \cdot U \cdot I \cdot \cos\varphi = k \cdot P. \qquad [4.20]$$

If this requirement is met, then the speed of the aluminum plate will be proportional to the power consumed:

$$\omega = k \cdot P \qquad [4.21]$$

The consumption of electricity is determined by differences in the dial readings at the beginning and the end of an interval. This requirement is not met in the beginning because Φ_I flux is not completely in phase with current I due to losses in the iron plate in the electromagnet, but lags behind in the amount of angle α. Also, the current of the voltage electromagnet lags behind the voltage with an angle smaller than 90° due to the resistance of voltage coil. To make the induction meter work properly, one flux Φ_U of the voltage electromagnet must lag behind the voltage U at an angle β (Figure 4.16), so that it follows:

$$\beta = \alpha + 90° \qquad [4.22]$$

Phase tuning is achieved in two parts. First tuning is done with an appropriate structure, so that only part of the flux passes through the aluminum plate, while fine tuning is done by turning the flux of the current electromagnet with the help of resistive coil of several turns.

The induction electricity meter connects to circuits just like wattmeters (Figure 4.17).

Regulations further insist that the meter must be connected to the main connector as closely as possible; the meter must be inaccessible to the customer, while the current electromagnet terminal must be connected to phase lead F, never to zero-lead 0.

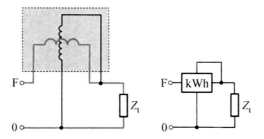

Figure 4.17 Connection of induction meters as in wattmeters

Special electricity meters allow the application of tariff policy for electricity distributors, with the aim of improving the efficiency of electricity production and consumption. Therefore, there are numerous additional features of meters, such as: multi tariff possibilities, meters with peak consumption information, meters with pulse encoders, etc. All these special features are made possible with modern electronic meters (Chapter 10.8).

4.5.2 Electricity Meter Testing

Electricity meters are tested using several methods:

a) The testing of meters is carried out by comparing consumption W_p, shown by the tested meter, with consumption W_s that was determined simultaneously by using a precise Watt-meter and clock. The testing can be done using the real load or with the help of artificial load (Fig. 4.18), in which the voltage and current branches are each connected to its source.

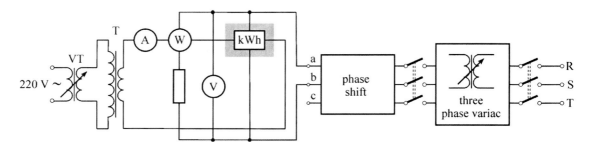

Figure 4.18 Testing of electricity meters with artificial load

The source connected to the voltage branch has only a small current, while the source connected to the current branch provides very little of the voltage that is required just to cover the voltage drops in the investigated branch and control device. In this way, the meter testing is enabled with small power consumption, which is important when the testing is done on a large series of meters. It is also possible to adjust voltages and currents, and phase shifts between them. Testing meters with the watt-meter and clock is done by comparing the tested meter with data obtained from a precise watt-meter and clock. The energy consumed is calculated as $W_s = Pt$, while the energy shown by electricity meters is calculated as:

$$W_p = \frac{N}{c} \cdot \frac{3600 \cdot 10^3}{\text{kWh}} \text{Ws} .\qquad [4.23]$$

where N is the number of revolutions the meter makes, and c is the meter constant.

The percentage error of the meter is then:

$$p = \frac{W_p - W_s}{W_s}\qquad [4.24]$$

b) Testing can be performed by comparison with the **precise electricity meter**. The current branches of tested and precise meters are connected in series and the voltage branches are connected in parallel. Comparison of the electricity meters is done at various loads. The precise meter is usually electronic, and the whole process of testing is automated, especially in manufacturer's testing stations and supervisory institutions.

c) The final meter test is performed after the adjustment, prior to the beginning of use, and is carried out to test the continuous operation to detect errors that could not be found in previous short-term tests.

Selected bibliography:

1. V. Bego, *"Measurements in Electrical Engineering"*, (in Croatian) Tehnička knjiga, Zagreb, 1990
2. D. Vujević, *"Measurements in Electrical Engineering – Laboratory Experiments"*, (in Croatian), Zagreb, 2001
3. IEC 62052 *Electricity Metering Equipment (AC) General Requirements, Tests and Test Conditions*, 2003
4. D. Vujević, B.Ferković, *"The Basics of the Electrotechnical Measurements, part I"*, Zagreb 1994
5. SACI, S.A. *Industrial Constructions, Analogue Instruments*, Catalogue, http://www.3ymf.ru/electro/saci/files/saci_analogicos.pdf, Acquired September 6th, 2010.
6. DIN 16257, *Nominal Positions and Symbols Used for Measuring Instruments*, Deutsches Institut Fur Normung E.V. (German National Standard) / 01-Mar-1987
7. DIN EN 60051-1, *Direct Acting Indicating Analogue Electrical-Measuring Instruments and Their Accessories - Part 1: Definitions and General Requirements Common to All Parts* (IEC 60051-1:1997); EN 60051-1:1998
8. Circutor, *Electrical measurements and control, Analogue Instruments*, Catalogue, http://img.icnea.net/Forum/E3032/ftp/Ca_M1_03.pdf, Acquired September 06, 2010.
9. Tyco electronics, *Analogue Instruments, Catalogue*, http://cromptonmeters.com/Crompton%20PDF/Analogue_Instruments_2006.pdf, Acquired September 06, 2010.

5. BRIDGES AND CALIBRATORS

Zero methods are indirect measurements that can be of the **bridge** or **compensation** type. Bridges are used to measure impedance, while the compensation method is used for comparison of a known voltage source with an unknown voltage source. Both compensation and bridge methods require the use of a sensitive instrument – the null detector. Measurement is performed by changing the known quantity until the null detector indicates "zero." This is why these methods are called *null methods*. Precise analogue instruments with moving coils and permanent magnets, as well as electronic instruments, are used as null detectors.

Analogue null detectors are sometimes called **galvanometers**. The **oscilloscope** is used as a null detector in AC measurements where not only amplitude, but also the phase, needs to be equalized. The most important characteristic of the null detector, concerning the accuracy, is its sensitivity.

5.1 DC BRIDGES

5.1.1 Wheatstone Bridge

The **Wheatstone** bridge is used to measure the resistances from 1 Ω to 1 GΩ. In this method, the null detector **N** is used to measure the difference between the voltage drops in two branches of the bridge (Figure 5.1).

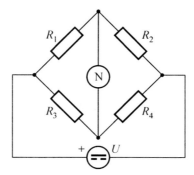

Figure 5.1 Wheatstone bridge

When the null detector indicates zero, the bridge is in **balance:** the same voltage is on resistors R_1 and R_3, and resistors R_2 and R_4, and the following applies: $R_1 \cdot R_4 = R_2 \cdot R_3$. If three resistances in the bridge are known and no current flows through the null detector, the fourth, unknown, resistance may be determined as:

$$R_1 = \frac{R_2 \cdot R_3}{R_4} \qquad [5.1]$$

As can be seen in equation [5.1], the voltage of the power source is irrelevant for the calculation of the unknown resistance. However, the higher the supply voltage, the greater the sensitivity of the null detector which leads to the possibility of increased measurement precision. However, higher voltages cause heating of the resistors and other adverse effects

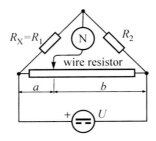

Figure 5.2 Wheatstone bridge with sliding wire

There are two basic versions of the Wheatstone bridge – one with **decade resistors** and one with a **sliding wire** (Figure 5.2). The commercial Wheatstone bridge with sliding wire is shown in Figure 5.3.

Figure 5.3 Commercial Wheatstone bridge

The Wheatstone bridges with decade resistors usually have sets of decades in the second branch of the bridge such as 10x0,1 Ω; 10x1 Ω; 10x10 Ω; 10x100 Ω; 10x1000 Ω, so in this case R_2 can get a value from 0,1 Ω to 11111 Ω. To enable easier calculation, the ratio R_3/R_4 is selected as 1/1000, 1/100, 1/10, 1, 10, 100, etc. Such a bridge can achieve limits of error less than ±0.02%.

In the case where the smallest change which can be achieved in the resistance R_2 is, for example, 0.1 Ω, it can be impossible to achieve zero deflection on the null detector because at one value of resistor R_2', the null detector will have a deflection α_1 on the one side, and with the value of resistor $R_2' + 0{,}1$ Ω, a deflection of α_2 on the other side. In this case, the value of R_2 at which zero deflection will be obtained can be calculated by linear interpolation:

$$R_2 = R_2' + \frac{\alpha_1}{\alpha_1 + \alpha_2} \cdot 0{,}1\,\Omega \qquad [5.2]$$

The Wheatstone bridge with **sliding wire** is smaller in size, but usually has larger error limits. The ratio R_3/R_4 is modified with changing the lengths **a** and **b** of the precisely calibrated sliding wire by moving the slider (Figure 5.2). As R_2 has a fixed value, the unknown resistance R_1 is calculated according to:

$$R_1 = R_2 \cdot \frac{a}{b} \qquad [5.3]$$

The accuracy of the measurement depends on the position of the slider. The most accuracy is achieved when the slider is closer to the middle of the wire; accuracy is worsened by moving the slider closer to the end of each wire.

5.1.2 Sensitivity of Wheatstone bridge

If the Wheatstone bridge is used to measure resistance R_X and if the limits of errors must not exceed $\pm \Delta R_X$, it is necessary that the change of resistance R_X by ΔR_X cause noticeable deflection of the null detector. Therefore, it is necessary to achieve satisfactory sensitivity of the bridge. For the sensitivity of the bridge, the following term is used:

$$\delta_{min}(\%) = \frac{\Delta R_{X\,min}}{R_X} 100\% \qquad [5.4]$$

It represents the smallest percentage change of the measured resistance or any resistance in the other branches of the bridge that can still be observed in the null detector. The minimal uncertainty of the Wheatstone bridge is a function of the **current constant** of the null instrument C_i, null detector resistance R_5, resistances in branches of the bridge, and voltage on the bridge:

$$\delta_{min} = \frac{C_i}{10U}\left[R_1 + R_2 + R_3 + R_4 + R_5\left(\frac{R_1}{R_2} + 2 + \frac{R_2}{R_1}\right)\right] \qquad [5.5]$$

It is important to emphasize that the uncertainty due to insensitivity of the bridge is only one component of measurement uncertainty when using the Wheatstone bridge.

5.1.3 Partially Balanced Wheatstone Bridge

Since the deflection of the null detector is directly related to the change of resistances in the Wheatstone bridge, a change of resistance in any branch can be continuously observed. If the resistance values of all branches in the bridge are $R_1=R_2=R_3=R_4=R$, and a resistance value in the fourth branch changes its value by ΔR, then the voltage in the diagonal of the bridge is:

$$U_d = U \frac{\frac{\Delta R}{R}}{4 + 2\left(\frac{\Delta R}{R}\right)} \qquad [5.6]$$

where U is the supply voltage of the bridge. Satisfactory linearity is achieved for the resistance changes of up to 10% in one branch of the bridge. The partially balanced Wheatstone bridge is widely used for measuring non-electric quantities such as temperature, using resistance measurement transducers or strain gauges.

5.1.4 Thompson Bridge

The double Kelvin or Thompson Bridge was discovered by **Lord Kelvin** in 1863, as the inadequacy of the Wheatstone bridge for measuring small resistance was realized (Figure 5.4).

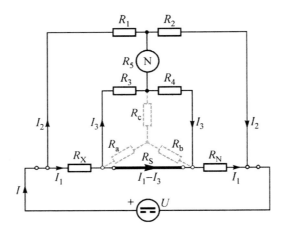

Figure 5.4 Thompson or double Kelvin bridge

With this bridge, the influence of resistance wires and contact resistances are removed. The "voltage" terminals of unknown resistor R_X and standard resistor R_N are connected to high-ohm resistances R_1, R_2, R_3, and R_4, to which the resistance wires and contact resistance are negligible. In the case of balance, R_X is calculated as:

$$R_X = R_N \frac{R_1}{R_2} + \frac{R_4 R_S}{R_3 + R_4 + R_S}\left(\frac{R_1}{R_2} - \frac{R_3}{R_4}\right) \qquad [5.7]$$

If R_1, R_2, R_3, and R_4 are chosen such as $\frac{R_1}{R_2} = \frac{R_3}{R_4} = n$, the above expression is simplified:

$$R_X = R_N \frac{R_1}{R_2} = R_N n \qquad [5.8]$$

5.2 AC WHEATSTONE BRIDGE

The AC Wheatstone Bridge (Figure 5.5) is an impedance bridge which consists of four impedances. By choosing appropriate elements in the bridge, it can be used to measure not only resistances, but also self-inductances, mutual inductances, capacitances, loss angles, and frequency.

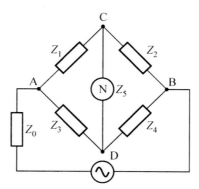

Figure 5.5 AC Wheatstone bridge

The basic requirement of equilibrium is similar to that of the DC bridge, but this time the balance equation is achieved using the complex quantities, and both the magnitude and phase must be equalized in order for the null detector to indicate "zero":

$$\boldsymbol{Z_1 Z_4 = Z_2 Z_3} \qquad [5.9]$$

If the impedance is written in the form $Z=R+jX$, the equation can be written as:

$$(R_1+jX_1)\cdot(R_4+jX_4) = (R_2+jX_2)\cdot(R_3+jX_3) \qquad [5.10]$$

The real and imaginary components of the equation must be equal, so the above equation can be broken down to two equations:

$$R_1 R_4 - X_1 X_4 = R_2 R_3 - X_2 X_3 \qquad [5.11]$$

$$R_1 X_4 + R_4 X_1 = R_2 X_3 + R_3 X_2. \qquad [5.12]$$

Using these two equations allows us to determine the real and imaginary components of the unknown impedance using the known real and imaginary components of other impedances.

When tuning the balance of the bridge, the voltage U_{CD} must be taken into account:

$$U_{CD} = U \frac{Z_1 Z_4 - Z_2 Z_3}{(Z_1 + Z_2)(Z_3 + Z_4)} \qquad [5.13]$$

By reducing voltage U_{CD} to zero, the balance of the bridge will be established. U_{CD} voltage will be equal to zero if the numerator will be equal to zero, and for the balance of the bridge, only the vector $z=Z_1Z_4-Z_2Z_3$ can be observed. This vector can be displayed in the form:

$$z = (R_1+jX_1)(R_4+jX_4)-(R_2+jX_2)(R_3+jX_3) \qquad [5.14]$$

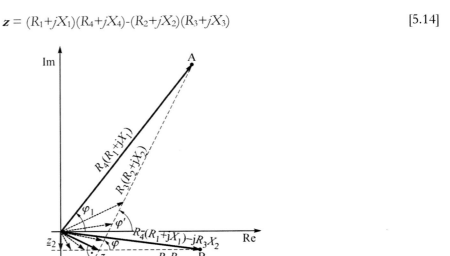

Figure 5.6 Balancing the AC Wheatstone Bridge

If balancing the bridge can be achieved in such a way that in the first tuning only the real part of the vector z is changed, and in the second tuning only the imaginary part of the vector z is changed, then the peak of the vector z will move in the complex plane in two lines, one parallel to the horizontal axis, and the other parallel to the vertical axis. Such an adjustment, where the lines of adjustments are perpendicular to each other, is called **independent adjustment** and can theoretically be achieved in only two attempts. However, in practice, the balancing is done in several steps, depending on the angle of two lines of adjustments (dashed lines in Figure 5.6). The request for practically independent balancing of the bridge leads to only a few basic combinations of AC bridges, some of which will be discussed in subsequent chapters. The null detector must be capable of detecting very small AC voltages. As stated before, the oscilloscope is often used as a null detector, and sometimes even the headphones.

5.3 DC COMPENSATION METHODS
There are two compensation methods: potentiometer and ammeter. As the ammeter compensation method is already outdated, only the potentiometer method will be described.

The principle of the potentiometer compensation procedure is shown in Figure 5.7. At the beginning of the measurement, the desired auxiliary current I_P is chosen by changing the resistor R_P. Then the auxiliary current, which is not changed anymore, flows through the AB potentiometer via resistance R_P.

BRIDGES AND CALIBRATORS

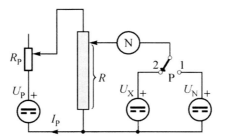

Figure 5.7 Compensation procedure

The switch P is then set to position 1, and the voltage standard output U_N is compared with the voltage drop on the potentiometer. The slider of the AB potentiometer is moved until the null detector reaches zero. Then, the following is true:

$$U_N = I_P \cdot R_{P1} \qquad [5.15]$$

Afterwards, switch P is set to position 2 and the compensation procedure is repeated for the unknown voltage U_X. The zero deflection on the null detector is reached for another resistance R_{P2}, and the following applies:

$$U_X = I_P \cdot R_{P2} \qquad [5.16]$$

Since the auxiliary current I_P did not change during measurements, the unknown voltage U_X can be calculated as:

$$U_X = U_N \frac{R_{P2}}{R_{P1}} \qquad [5.17]$$

The auxiliary current does not have to be known, and the procedure can be further simplified by choosing the appropriate values for U_N and R_{P1}, so the determination of an unknown voltage U_X is reduced to only one measurement with a switch in the position 2. For example, for $U_N = 1$ V and $R_{P1} = 10000$ Ω the unknown voltage U_X is:

$$U_X = 10^{-4} \cdot R_{P2} \qquad [5.18]$$

The accuracy of the compensation procedure depends on the quality of the compensating resistor. The compensating resistor can be made of several good resistor decades, or a combination of resistor decades and helicoidal potentiometers.

The advantage of the compensation procedure in relation to measuring voltage using the voltmeter is that at the time of compensation, the unknown voltage source is unloaded, so there will be no voltage drop on its internal resistance.

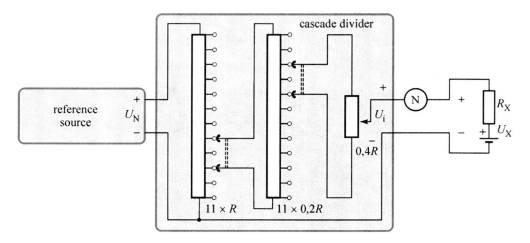

Figure 5.8 Compensator with cascade divider

Over the years, various compensators have been used, but nowadays, cascade compensators with the **Kelvin-Varley** divider (Figure 5.8) are popular. The main purpose of the Kelvin-Varley divider is to divide the accurately known input voltage U_N. Two decades of A and B (11xR and 11x0.2R) have a switch with two mutually insulated sliding contacts, which are interconnected and separated by two steps. The Helicoidal potentiometer (0.4 R) is connected in parallel to the two decade resistors B (which are connected in series and also have the same resistance 2x0.2R = 0.4R). Such a parallel combination has a total resistance of 0.2R. The total resistance of decade B is therefore 10x0.2R=R, which is connected in parallel to two resistances of decade A (2xR=2R). The total resistance of decade A is then 10xR. Thus, regardless of the decade or potentiometer slider position, the resistance which is connected to the reference source always remains the same (10xR), ensuring a constant auxiliary current. The output voltage U_i is determined by the input voltage and the potentiometer decade switch positions. The unknown voltage U_X is determined by comparison with the output of the voltage divider U_i using the null-instrument N. The double slider solves the problem of thermal EMF of contacts, because they are, in the loop of each decade, a serial connection with the opposite polarity. Number of decades depends on the required accuracy, and is used for accurate measurement up to eight decades.

5.4 CALIBRATORS

Calibrators are the **sources** of fixed and accurately known voltages and currents, which may be changed in small steps (the best calibrators can adjust the voltage in increments of 10nV and 10nA), and are used for calibration of measuring instruments. They are divided into AC and DC calibrators. AC calibrators enable frequency adjustments in increments of several hertz. Today, multifunction calibrators are manufactured with multiple output quantities such as current, voltage, AC and DC, and even resistance. They also have a GPIB interface for connecting to a computer (see Chapter 11), thereby enabling the automatic calibration of measuring instruments.

5.4.1 DC Calibrator

The DC calibrator consists of an accurately known voltage source (obtained using sources with the Zener diode), voltage divider, and amplifier. In older calibrators, resistance dividers were used, but today calibrators adjust the output voltage using the **analogue to digital converter with pulse** width modulation - PWMDAC. A block diagram of a modern DC voltage calibrator is shown in Figure 5.9.

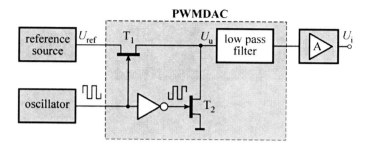

Figure 5.9 DC voltage calibrator basic circuit

The reference voltage is forwarded to PWMDAC, which contains two transistors, T1 and T2, excited from the silica-controlled oscillators and microprocessor to act as switches that turn on and off, so the low pass filter receives either reference voltage or zero (Figure 5.10). The ratio of pulse duration t to the total period T can be varied in wide ranges, so the output voltage of the low pass filter is accurately known:

$$U_i = U_{ref} \frac{t}{T} \qquad [5.19]$$

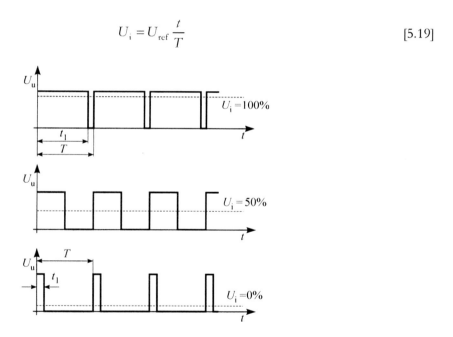

Figure 5.10 The PWMDAC principle of work

The output voltage is amplified in several stages, so the output ranges are from 220 mV to 1100 V. The commercial Fluke 5500 type calibrator is shown in the Primary Electromagnetic Laboratory of Croatia.

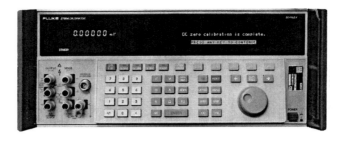

Figure 5.11 Precise DC voltage calibrator Fluke 5500 in the Primary Electromagnetic Laboratory of Croatia

The **current calibrator** works on the same principle, only the voltage produced at the exit of the PWMDAC is amplified with current output amplifiers.

5.4.2 AC Calibrator

The AC voltage is produced by sine oscillators. To achieve the narrow error limits of output sinusoidal voltage, its effective value is compared with the highly accurate DC voltage supplied by PWMDAC, using the circuit with thermocouples. The output voltage of thermocouples is proportional to the difference between the effective values of sine and DC voltage and is used in feedback mode to adjust the output voltage of the oscillator. Tuning is done until the effective value of sine and DC voltage are equal, in which case the output voltage of the thermocouple will be zero (Figure 5.12).

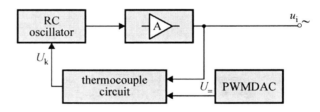

Figure 5.12 AC calibrator's principle of work

Selected bibliography:
1 V. Bego, "*Measurements in Electrical Engineering*", (in Croatian) Tehnička knjiga, Zagreb, 1990
2 D. Vujević, B.Ferković, "*The Basics of the Electrotechnical Measurements*", part I, Zagreb 1994
3 *AC Bridge Circuits*, http://www.allaboutcircuits.com/vol_2/chpt_12/5.html, Acquired September 19th, 2010.
4 D. Vujević, "*Measurements in Electrical Engineering – Laboratory Experiments*", (in Croatian), Zagreb, 2001
5 R. Ince, D. Atesalp: "*Wheatstone Bridge for Precision Automated Calibration of High Resistance Standards*", XIV IMEKO World Congress, Proceedings, Volume IVA, pp., 1997.
6 H. Sasaki, N. Nishinaka, S. Shida: "*High Precision Automated Resistance Measurement Using a Modified Wheatstone Bridge*", CPEM '88 Digest, pp. 157-159.
7 D. Braudaway: "*Precision Resistors: A Review of Techniques of Measurement, Advantages, Disadvantages*", IEEE Trans. Instrum. Meas. Vol. 48, No. 5, pp. 884-888, October 1999.
8 J. L. Thomas: "*Precision Resistors and Their Measurement*", Natl. Bur. Stand. (U.S.), Circular 470, October 1948.

9   *Bridge Circuits*, http://www.allaboutcircuits.com/vol_1/chpt_8/10.html, Acquired September 19th, 2010.
10 A. F. Dunn: "*Calibration of a Kelvin–Varley Voltage Divider*", IEEE Trans. Instrum. Meas., pp. 129-139, June- Sept. 1964.
11 "*Calibrating the Hewlett Packard 3458A DMM with the Fluke 5720A Multifunction Calibrator*", Fluke Application Note, http://www.fluke.com/scripts/calibrators/appnotes, Retrieved October 10th, 2010
12 "*Calibration: Philosophy in Practice*", Fluke, Second Edition, 1994.

6. INSTRUMENT TRANSFORMERS

Instrument transformers are used to measure high AC voltages, large currents, power, and energy, and also to extend the measurement ranges of measurement instruments. Their purpose is to galvanically separate high voltage circuits from low voltage circuits. Voltage transformers are used to expand the range of voltage measurement, whereas current transformers are used to expand the range of current measuring instruments. Transformers reduce the voltage in a known ratio to values that are appropriate for measurement: the currents are reduced to 1 A or 5 A, and voltages are reduced to 100 V and 200 V for line voltages, or to $\frac{100}{\sqrt{3}}$ and $\frac{200}{\sqrt{3}}$ for phase voltages. Transformers consist of a core made of magnetic materials, and primary and secondary windings.

6.1 CONNECTING INSTRUMENT TRANSFORMERS

The voltage transformer is connected in **parallel** with the loads whose voltage needs to be measured (Figure 6.1). The current through the primary winding must be considerably smaller than the current through the load, just like the current through the voltmeter must be as low as possible in direct voltage measurement. The primary winding of the current instrument transformer is connected in **series** with the load, and the voltage drop on the primary winding must be much smaller than the voltage drop on the load, similar to the voltage drop on ammeters in direct current measurements.

Figure 6.1 Connection of voltage and current transformers

6.2 IDEAL AND REAL TRANSFORMERS

An ideal transformer transforms the primary voltage and current in a constant ratio, i.e., it is proportional to the ratio of the number of windings. There is also no phase shift between the primary and secondary voltage and current.

Thus, the following applies for the ideal voltage transformer:

$$\frac{U_1}{U_2} = \frac{N_1}{N_2} \qquad [6.1]$$

For the ideal current transformer, the formula is similar:

$$\frac{I_1}{I_2} = \frac{N_2}{N_1} \qquad [6.2]$$

An ideal transformer has no resistance of primary and secondary windings, and has a magnetic core of infinite conductivity, which does not require the magnetization current, so there are no leakage core fluxes (Figure 6.2).

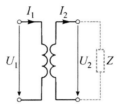

Figure 6.2 Equivalent circuit of ideal transformer

In reality, there are several causes of errors in measurements with instrument transformers (Figure 6.3). The primary winding has resistance R_1, while a part of the magnetic flux covers only the primary winding. Its influence can be described by leakage inductance $L_{1\sigma}$, or reactance $X_1 = \omega L_{1\sigma}$. In analogy to the primary winding, the secondary winding has resistance R_2 and leakage inductance $L_{2\sigma}$, or reactance $X_2 = \omega L_{2\sigma}$. For maintaining the mutual flux of the core, magnetizing current I_0 is required, which causes losses in the core. Active and reactive core losses are described with the resistance R_0 and reactance X_0, connected in parallel to the ideal transformer.

The above are all elements of the transformer equivalent circuit shown in Figure 6.3. It can be simplified if the values of primary voltage, current, and resistance are converted to secondary values and if the ideal transformer is omitted:

$$U_1^{''} = U_1 \frac{N_2}{N_1} \,;\, I_1^{''} = I_1 \frac{N_1}{N_2} \,;\, R_1^{''} = R_1 \left(\frac{N_2}{N_1}\right)^2 \,;\, X_1^{''} = X_1 \left(\frac{N_2}{N_1}\right)^2 ;$$

$$I_0^{''} = I_0 \frac{N_1}{N_2} \,;\, I_g^{''} = I_g \frac{N_1}{N_2} \,;\, I_\mu^{''} = I_\mu \frac{N_1}{N_2} \,;\, X_0^{''} = X_0 \left(\frac{N_2}{N_1}\right)^2 ;$$

[6.3]

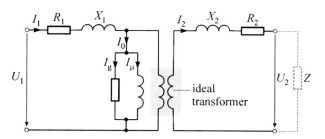

Figure 6.3 Equivalent circuit of real transformer

6.3 VOLTAGE INSTRUMENT TRANSFORMER

The ratio between the voltages U_1 and U_2 of the real transformer does not equal the actual transformer turns ratio of N_1 and N_2, and they are also not in phase (Figure 6.4). The reason is a voltage drop created by the primary current I_1 on the resistance R_1 and reactance X_1 and the voltage drop created by the secondary current I_2 on the resistance R_2 and reactance X_2. These voltage levels are lower and closer to the ideal transformer if the transformer is less loaded, so the losses vary with load and can go from no-load to full-load losses. The vector diagram of the instrument voltage transformer is shown in Figure 6.5, where it is possible to observe the errors quantitatively. For greater transparency, voltage and phase errors are drawn larger than they actually are.

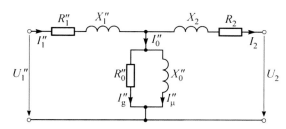

Figure 6.4 Equivalent circuit of real voltage transformer without the ideal transformer

Description of Figure 6.4: the main magnetic flux Φ induces voltage U_i'' in primary and secondary windings. This vector voltage is perpendicular to the magnetic flux vector. Through secondary windings, where instruments and loads are connected, current I_2 will flow. This current lags from voltage U_2 for some angle β_2 and will produce voltages $R_2 I_2$ and $jX_2 I_2$. Therefore, the voltage U_2 is smaller than the induced voltage U_i''. Due to voltage drops on $R_1 I_1$ and $jX_1 I_1$ caused by current I_1 in primary windings, the voltage applied to primary windings U_1'' is greater than the induced voltage U_i''. For maintaining the flux Φ current I_0'' is necessary. Its reactive component I_μ'' is in phase with the magnetic flux of the core, while component I_g'' is 90° phase-shifted and takes care of the losses in the iron core. The current I_1'' is equal to the vector sum of secondary current I_2 and magnetizing current I_0''. As is evident from the vector diagram in Figure 6.5, there is a phase shift δ between the voltages U_1'' and U_2 that represents the **phase error** of the voltage instrument transformer, usually expressed in minutes (δ'). The same holds for the **voltage error** of the transformer p_n:

$$p_n = \frac{K_n U_2 - U_1}{U_1} \cdot 100\% \qquad [6.4]$$

where $K_n = U_{1n}/U_{2n}$ is the ratio between the nominal primary and secondary voltage.

International standards (IEC 60044-2) divide voltage transformers into five classes of accuracy (Table 6.1.) which must be met for voltages between 80 and 120% of rated voltage and loads between 25 and 100% of rated load.

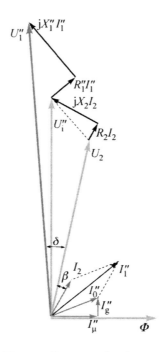

Figure 6.5 Vector diagram of voltage transformer

Table 6.1. Accuracy class of voltage instrument transformers

Accuracy class	Errors	
	voltage (%)	phase (minutes)
0,1	± 0,1	± 5
0,2	± 0,2	± 10
0,5	± 0,5	± 20
1	± 1	± 40
3	± 1	not defined

The same standard also regulates 12 rated burdens for voltage transformers, and manufacturers are encouraged to use six, which are printed in bold:

10 VA, 15 VA, **25 VA**, 30 VA, **50 VA**, 75 VA, **100 VA**
150 VA, 200 VA, 300 VA, 400 VA, **500 VA**

Terminals of primary instrument transformers are referred to in capital letters **U** and **V**, and secondary terminals are referred to in small letters **u** and **v**. If the voltage transformer is designed for connection to phase voltage, then primary terminals are marked with **U** and **X**, and secondary with **u** and **v**. Three-phase voltage instrument transformers are marked with **U, V, W** for the phases and **X** for the neutral for primary terminals, and **u, v, w,** and **x** for secondary terminals.

For lower voltages, transformers are manufactured as a dry construction insulated with epoxy resin. For voltages of more than a dozen kilovolts, they are made with the winding and core immersed in a cauldron with transformer oil.

6.4 CAPACITIVE MEASURING TRANSFORMERS

To measure high voltages above 110 kV, capacitive transformers are used. The main part of such a transformer is a high capacity capacitor C_1, connected in series with a low voltage capacitor C_2 which is connected to the voltmeter (Fig. 6.6).

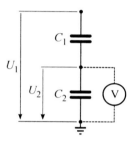

Figure 6.6 Capacitive transformer

If the voltmeter has negligible consumption, then the measured voltage is obtained as:

$$U_1 = U_2 \cdot \left(1 + \frac{C_2}{C_1}\right). \tag{6.5}$$

If the voltmeter internal resistance cannot be ignored, then the measured voltage is:

$$U_1 = U_2 \cdot \sqrt{\left(1 + \frac{C_2}{C_1}\right)^2 + \left(\frac{1}{\omega C_1 R_V}\right)^2} \tag{6.6}$$

When using capacitive instrument transformers in high voltage applications, the voltmeter is not the only instrument to be connected, as there are also wattmeters, electricity meters, and other instruments: their consumption can reach over 100 VA, which significantly affects the voltage ratio. The solution is then to reduce the value of the capacitor C_2, which increases the voltage drop on this capacitor to several kilovolts. Then, this voltage is measured through a conventional transformer (Figure 6.7).

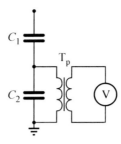

Figure 6.7 Capacitive transformer with conventional transformer connected to C_2

This setup can be additionally improved by adding a coil with inductance L ahead of the transformer T_p (Figure 6.8). The input impedance between the points 1 and 2 is:

$$Z_u = \frac{1}{\varpi(C_1 + C_2)} \qquad [6.7]$$

The capacitor C_1 can be considered as parallel to the capacitor C_2 due to small impedance of the network. As Z_u is capacitive, it can be significantly lowered with the addition of inductance L, so it is:

$$L\varpi = \frac{1}{\varpi(C_1 + C_2)} \qquad [6.8]$$

The impedance of this combination does not change the voltage, which is desirable.

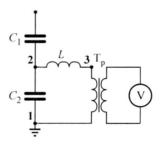

Figure 6.8 Capacitive transformer with the coil added ahead of transformer Tp

6.5 CURRENT INSTRUMENT TRANSFORMER

The ratio between currents I_1 and I_2 of the real transformer is not equal to the actual transformer turns ratio of N_2 and N_1. These currents are not in phase because the flow of the current needs a voltage induced in the secondary winding. To be induced, the voltage needs a certain number of primary windings, which are not compensated by the secondary winding, but are used for magnetization of the core. The equivalent circuit of the current transformer is seen in Figure 6.9.

INSTRUMENT TRANSFORMERS

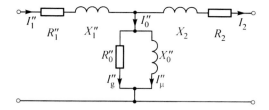

Figure 6.9 Current transformer equivalent circuit

The windings are the cause of current and phase transformer errors, so it is important to keep the magnetizing current I_0 as low as possible. Therefore, materials with high permeability are used for cores. For the same reason, the internal resistance of measuring instruments connected to current transformers must have a relatively low value. If the terminals of the current instrument were left open, then the entire primary current would magnetize the core, which would cause extreme saturation and high voltage at the secondary terminals. High induction causes high losses in the iron and leads to excessive heat, and even to the destruction of the transformer. For this reason, the secondary terminals in operation must never remain open. The errors of the current instrument transformers can be quantitatively observed in the vector diagram of the instrument current transformer as shown in Figure 6.10.

The main magnetic flux Φ induced in the primary and secondary winding voltage is U_i''. This voltage is perpendicular to the magnetic flux. Through the secondary winding current I_2 flows, which lags the voltage U_2 for some angle β_2 and generates the voltage drops $R_2 I_2$ and $j X_2 I_2$. Because of these voltage drops, voltage U_2 is smaller than the induced voltage U_i''. To maintain the flux Φ, current I_0'' is required. Its reactive component I_μ'' is in phase with the flux of the magnetic core, while the active component I_g'' is in phase with the induced voltage U_i'', and covers the losses in iron.

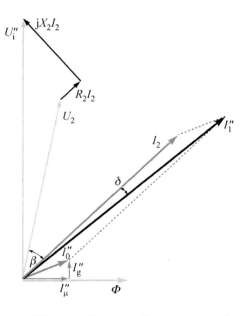

Figure 6.10 Vector diagram of current transformer

The vector sum of the currents I_2 and I_0'' is equal to the primary current I_1'', which is converted to the secondary side of the transformer. Currents I_2 and I_1'' are out of phase by angle δ, which represents the **phase error** (expressed in minutes) and by magnitude, which can be calculated according to the following expression:

$$p_i = \frac{K_n I_2 - I_1}{I_1} \cdot 100\%. \qquad [6.9]$$

where $K_n = I_{1n}/I_{2n}$ is the nominal ratio of the current instrument transformer. One of the current transformers at the Faculty of Electrical Engineering and Computing in Zagreb is shown in Figure 6.11. Figure 6.12 shows a current transformer with two cores and two secondary windings.

Figure 6.11 Current transformer at the Faculty of Electrical Engineering and Computing in Zagreb

Table 6.2 Error limits for current instrument transformer

Accuracy class	Current errors (%):				Phase errors ('):			
	$0{,}05\,I_n$	$0{,}2\,I_n$	$1\,I_n$	$1{,}2\,I_n$	$0{,}05\,I_n$	$0{,}2\,I_n$	$1\,I_n$	$1{,}2\,I_n$
0,1	0,4	0,2	0,1	0,1	15	8	5	5
0,2	0,75	0,35	0,2	0,2	30	15	10	10
0,5	1,5	0,75	0,5	0,5	90	45	30	30
1	3,0	1,5	1,0	1,0	180	90	60	60
3	from $0{,}5\,I_n$ to $1{,}2\,I_n$				not defined			
5	from $0{,}5\,I_n$ to $1{,}2\,I_n$				not defined			

Nominal impedance Z_n is calculated from data on the nominal output load and the rated secondary current I_{2n}. Limits of error for current instrument transformers are given in Table 6.2; these limits must not be exceeded for transformers with the accuracy class 0.1, 0.2, 0.5, and 1 when they are loaded with power between 25% and 100% of the rated output load, and with a power factor of 0.8. The standard values of the current instrument transformer burden are:

2,5 VA; 5 VA; 10 VA; 15 VA; 30 VA

Terminals of the primary current transformer are referred to in capital letters **K** and **L**, and secondary terminals are referred to as lower case **k** and **l**.

INSTRUMENT TRANSFORMERS

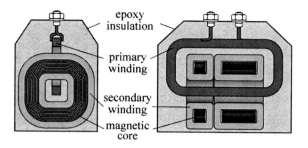

Figure 6.12 Current transformer with two cores and two secondary windings (cross section)

6.6 CURRENT INSTRUMENT TRANSFORMER ACCURACY TESTING

Current transformers are usually tested for accuracy according to two known methods. One is called the Schering and Alberti method and the other is according to Hohle.

6.6.1 Schering and Alberti Method

The Schering and Alberti method is shown in Figure 6.13.

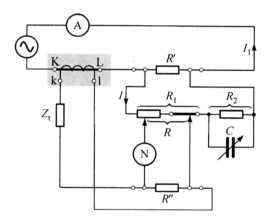

Figure 6.13 Schering and Alberti method for determining the current and phase error of the current transformers

The primary current of the tested current transformer (shaded) is flowing through the precise resistor R'. The primary current is compensated by the voltage drop of the second precise resistor R", through which a secondary current is flowing. To determine the current and phase error, the compensator with variable resistor R_1, variable capacitor C, and resistor R_2 is used. The null detector can be an oscilloscope or very sensitive electromechanical meter. The ratio of the primary and secondary current can be determined as:

$$K_m = \frac{I_1}{I_2} = \frac{(R_1 + R_2)R''}{RR'} \qquad [6.10]$$

The current error of the measuring transformer according to [6.9] can be rewritten as:

$$p_i = \frac{K_n - K_m}{K_m} \cdot 100\%. \qquad [6.11]$$

where $K_n = I_{1n}/I_{2n}$ is the nominal ratio of the current instrument transformer. The phase error is calculated as:

$$\delta \approx tg\delta = \frac{\omega R_2 C}{R_1 + R_2}. \qquad [6.12.]$$

6.6.2 Hohle Method

The Hohle method is shown in Figure 6.14. The primary windings of the standard transformer N and the transformer under test X are connected in series. The secondary windings are also connected in series. A difference of secondary currents I_{2X} and I_{2N}, designated as I_0, flows through the precise resistor R of small value.

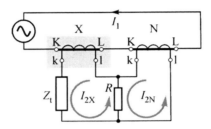

Figure 6.14 Hohle method

As the standard transformer error is negligible compared to the test transformer, the I_0 current is actually the total error of the test transformer. The current and phase errors (Figure 6.15) are determined from the components of current I_0, in which one component is perpendicular and the other is parallel to the current I_{2N}. The $\overline{OM}$ component which is in phase with current I_{2N}, is the absolute current error, while component $\overline{MN}$ is the phase error of the tested transformer.

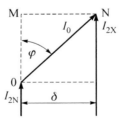

Figure 6.15 Vector diagram of current and phase errors according to the Hohle method

The voltage on resistor R caused by current I_0 is compensated by two mutually perpendicular voltages, one which is in phase with current I_{2N}. These voltages are generated by an auxiliary current transformer T_S and transformer T_M with mutual inductance M (Figure 6.16).

INSTRUMENT TRANSFORMERS

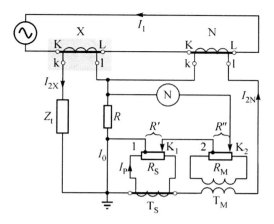

Figure 6.16 Compensation of I_0 current in the Hohle method

The secondary current I_p of the transformer T_S is proportional and in phase with current I_{2N} ($I_p=nI_{2N}$), and therefore the voltage drop on resistor R_S is also in phase with current I_{2N}. The voltage induced in the secondary windings of transformer T_M is phase-shifted from current I_{2N} by 90°; therefore, the voltage on resistor R_M is also shifted by 90°. After the compensation is achieved by moving the sliders K_1 and K_2, the voltage drop on resistor R is equal to the voltage drops on resistances R' and R''. The current error can be calculated as:

$$p_i = \frac{I_0 \cos\varphi}{I_{2N}} 100\% = \frac{nR'}{R_M} 100\% . \qquad [6.13]$$

The phase error is:

$$\delta \approx \frac{I_0 \sin\varphi}{I_{2N}} = \frac{R''M\omega}{RR_M} . \qquad [6.14]$$

6.6 WINDING CONFIGURATIONS

A single primary and secondary coil as discussed above is not the only possible configuration of transformers. Transformers do not have to be made with only two sets of windings - they can have several sets of windings.

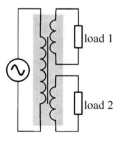

Figure 6.13 Transformer with one primary and two secondary windings

In Figure 6.13, three coils share a common magnetic core, and the relationships of winding turn ratios and voltage turn ratios are still valid. In fact, one of the secondary windings can be a step-up transformer, while the other can be a step-down transformer, and all are still electrically isolated from each other. A current transformer with two cores and two secondary windings can be seen in Figure 6.12.

Variable transformers (Figure 6.14) are capable of providing variable output voltage from zero volts to maximum voltage. This is achieved with a sliding contact which can be moved along the length of an exposed secondary winding.

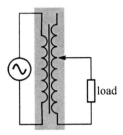

Figure 6.14 Variable transformer

Selected bibliography:
1. V.Bego, "*Measurements in Electrical Engineering*", (in Croatian) Tehnička knjiga, Zagreb, 1990
2. IEC 60044 – *Instrument Transformers*, International Electrotechnical Commision Standard, 2003
3. V. Bego: "*Instrument Transformers*", (in Croatian), Tehnička njiga, Zagreb, 1985
4. D. Vujević, "*Measurements in Electrical Engineering – Laboratory Experiments*", (in Croatian), Zagreb, 2001
5. D. Vujević, B.Ferković, "*The Basics of the Electrotechnical Measurements, Part I*", Zagreb 1994
6. Jason Starck, "*Transformers*", http://www.opamp-electronics.com/tutorials/ac_theory_ch_009.htm, Retrieved October 23, 2010
7. *Instrument Transformers*, http://www.sayedsaad.com/Protection/files/ VT_CT/1_VT_ CT.htm, Retrieved September 28th, 2010
8. IEEE std. C57.13-1993, *IEEE Standard Requirements for Instrument Transformers*, 1993

7. AMPLIFIERS IN MEASUREMENT TECHNOLOGY

Electronic circuits can be used to enhance the properties of measuring instruments and expand their field of application. Electronic devices may be analogue or digital, and measuring amplifiers are an important part of analogue electronic instruments.

7.1. MEASURING AMPLIFIERS

Measuring amplifiers have dual uses:

a) amplifying the low value of the measured quantity to a value that can be measured
b) adjusting the impedance to a desired value.

In special circuits, amplifiers are used to perform **mathematical operations** such as addition, subtraction, multiplication, division, integration, and derivation. Measuring amplifiers must have:

- high input impedance when used in voltage and resistance measuring instruments, and small impedance when used in current measuring instruments,
- constant amplification in a wide frequency band,
- uniformity of output depending on the input signal.

The main feature of the amplifier is its gain A_0, which is defined by the ratio of output and input signals. The output and input signals can be either current or voltage; thus, four combinations are possible:

1) V/V voltage amplifier
2) A/V transimpedance amplifier
3) V/A transconductance amplifier
4) A/A current amplifier

Voltage and current amplifiers are dimensionless, while the transconductance amplifier has the dimension of conductivity - Siemens - and the transimpedance amplifier has the dimension of resistance - Ohm. The gain of voltage, current, and power amplifiers is expressed in decibels (dB) – a logarithmic unit for expressing the ratio. According to the definition of decibel, it applies:

voltage n dB = $20 \log U_1/U_2$
current n dB = $20 \log I_1/I_2$
power n dB = $10 \log P_1/P_2$

The amplifier is represented with a symbol (Figure 7.1).

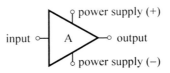

Figure 7.1 Amplifier symbol

The frequency range within which the amplifier has a fixed gain is called the frequency band; it is bound by the lower f_d and upper f_g cutoff frequencies. The upper and lower limit frequencies are determined at the points where the gain is decreased to 70.7% (-3 dB) of the primary amplification in the middle band (Figure 7.2).

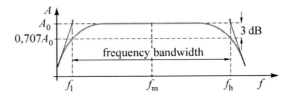

Figure 7.2 Frequency bandwidth of amplifier

Due to various influences (temperature, aging of components, changing the voltage, etc.) amplification is not constant, which is one of the conditions for the amplifier to be used for measuring purposes. Constant gain is achieved by the **negative feedback circuit (β)**.

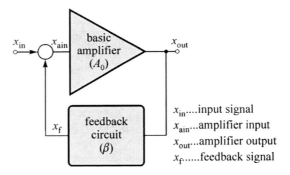

Figure 7.3 Negative feedback circuit

The principle of negative feedback voltage is shown in Figure 7.3. Part of the output signal is fed to the amplifier input, and at the input of the amplifier the signal is not x_{in}, but:

$$x_{ain} = x_{in} - x_f = x_{in} - \beta x_{out} \qquad [7.1]$$

where β is the factor of negative feedback. If the amplifier has a gain $A_0 = \dfrac{x_{out}}{x_{in}}$, ($A_0$ is called amplification of an open-loop), then it can be written:

$$x_{ain} = x_{in} - \beta A_0 x_{ain} \qquad [7.2]$$

or:

$$x_{ain} = \dfrac{x_{in}}{1 + \beta A_0}. \qquad [7.3]$$

The output signal is then:

$$x_{out} = A_0 x_{ain} = \dfrac{A_0}{1 + \beta A_0} \cdot x_{in}. \qquad [7.4]$$

The amplifier gain with feedback is:

$$A_f = \dfrac{A_0}{1 + \beta A_0}. \qquad [7.5]$$

If the product βA_0 is selected such that $\beta A_0 \gg 1$, then the gain of the amplifier with the feedback will be $A_f \approx \dfrac{1}{\beta}$, so that it practically does not depend on open-loop amplification A_0.

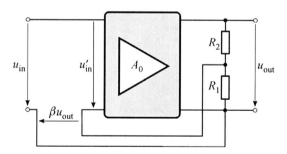

Figure 7.4 Negative feedback with resistor divider

If a resistance divider is chosen in branch β (Figure 7.4) to be:

$$\beta = \dfrac{R_2}{R_1 + R_2} \qquad [7.6]$$

then the gain of the amplifier with feedback is:

$$A_f \approx \frac{1}{\beta} = \frac{R_1 + R_2}{R_2} \quad [7.7]$$

Resistors R_1 and R_2 can be precision resistors, and so the gain A_f can also be very accurate. For example, if $A_0=10000$, and $\beta=0.099$, then the amplifier with negative feedback has the gain $A_f=10.101$, which is about 1000 times smaller than A_0. If for some reasons A_0 is reduced by 30% ($A_0 = 7000$), it will reduce the A_f only by 0.144 % and the amplification with feedback will be $A_f = 10.086$.

Negative feedback is also used to extend the frequency band of the amplifier (Fig. 7.5), as the amplification gain and cutoff frequency is constant, so that:

$$A_0 f_g = A_f f_{gf}. \quad [7.8]$$

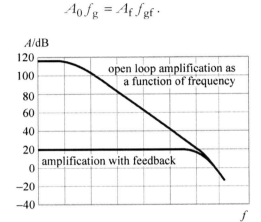

Figure 7.5 Open loop amplification as a function of frequency

Example: the amplifier gain without feedback is $A_0=10000$, and its upper limit frequency is $f_g=10$ Hz. Using negative feedback, the amplification is reduced to $A_f =100$ and thereby the frequency band is increased by:

$$f_{gf} = \frac{A_0 f_g}{A_f} = 1000 \text{ Hz}. \quad [7.9]$$

7.2. OPERATIONAL AMPLIFIERS

Operational amplifiers are mostly used as measuring amplifiers, and their symbol is a triangle (Figure 7.6).

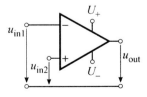

Figure 7.6 Operation amplifier inputs and outputs

Operational amplifiers are essentially voltage amplifiers with high open loop gain A_0, and two inputs and one output that are capable of accomplishing many different functional relationships (operations) between the inputs and outputs. Elements for a negative feedback are usually found outside of the amplifier. Two inputs are called **inverting** (marked with -) and **non-inverting** (marked with +). The term "inverting" is used because the input voltage u_{in1} brought to the inverting terminal is amplified and rotated by 180°, while the input voltage u_{in2} brought to the non-inverting input is only amplified. Table 7.1 shows the characteristics of ideal and real operational amplifiers. Real operational amplifiers draw a small current from each of their inputs due to bias currents and leakage. These currents flow through the resistances connected to the inputs and produce small voltage drops across those resistances - which must be taken in consideration in high precision DC applications. In AC signal applications, this seldom matters. The leakages can be reduced if voltage drops are equal in both inputs. Also, the accuracy can be improved if the DC input resistances of each input are matched. Power supply imperfections (e.g., power signal ripple, non-zero source impedance) may also lead to noticeable deviations from ideal operational amplifier behavior.

Table 7.1. Characteristics of ideal and real operational amplifiers

Characteristic	IDEAL	REAL
Input impedance	∞	$10^5 - 10^{12}$ Ω
Output impedance	0	25 – 50 Ω
Frequency band (open-loop)	∞	10 Hz
Gain	∞	$>10^5$

7.3 OPERATIONAL AMPLIFIER APPLICATIONS

The operational amplifier is used for many different applications in measurement technology, but in other technologies as well. In the following chapters, the most commonly-used applications will be presented. The formulas given to describe these applications assume that the ideal op-amp is used. If op-amps are considered for precise measurement applications, all the influencing factors and non-ideal characteristics must be taken into consideration.

7.3.1 Inverting Amplifier

Figure 7.7 shows the principle scheme of the inverting amplifier.

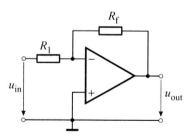

Figure 7.7 Inverting amplifier

The output voltage according to [7.1-7.5] is:

$$u_{out} = -u_{in}\frac{R_f}{R_1} \qquad [7.10]$$

7.3.2 Summing Amplifier

From inverting amplifiers, it is easy to make the summing amplifier (Figure 7.8):

$$u_{out} = -R_f \left(\frac{u_{in1}}{R_1} + \frac{u_{in2}}{R_2} + ... + \frac{u_{inn}}{R_n} \right) \quad [7.11]$$

If $R_1 = R_2 = ... R_n = R_f$, then the output voltage is the sum of input voltages:

$$u_{out} = -\sum_{k=1}^{n} u_{ink} \quad [7.12]$$

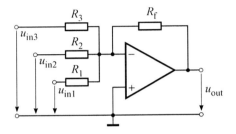

Figure 7.8 Summing amplifier

7.3.3 Non-Inverting Amplifier

The non-inverting amplifier is shown in Figure 7.9. The output voltage is:

$$u_{out} = u_{in} \cdot \left(1 + \frac{R_f}{R_1} \right) \quad [7.13]$$

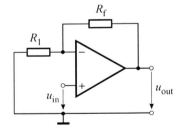

Figure 7.9 Non-inverting amplifier

It is characteristic of both types of amplifiers that the accuracy and constant gain depends only on the accuracy and stability of resistors R_1 and R_f.

7.3.4 Integrating Amplifier

The **integrating** amplifier is shown in Figure 7.10. Instead of a resistor in the feedback loop, there is a capacitor. The potential difference between inverting and non-inverting inputs is practically equal to zero and the junction of capacitor C and resistor R is the so-called "virtual zero." Therefore, the output voltage equals the voltage on the capacitor, and so it is:

$$u_{out} = \frac{1}{C}\int i_f dt = -\frac{1}{RC}\int u_{in} dt. \quad [7.14]$$

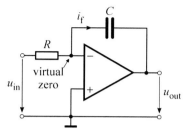

Figure 7.10 Integrating amplifier

As $\int i_f dt = \int i_{in} dt$ is actually the incoming charge, integrating amplifiers can be used to measure charge; such instruments are called **coulonmeters**.

7.3.5 Differentiator Amplifier

The differentiator amplifier is shown in Figure 7.11.

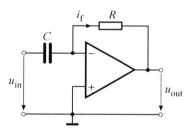

Figure 7.11 Differentiator amplifier

Unlike in the integrating amplifier, there is a resistor in the feedback loop, while a capacitor is placed at the inverting input terminal. The output voltage is obtained as follows:

$$u_{out} = Ri_f = -RC\frac{du_{in}}{dt}. \quad [7.15]$$

7.3.6 Logarithmic Amplifier

In the logarithmic amplifier, there is a transistor in the feedback loop (Figure 7.12). The output voltage is equal to the logarithm of the input:

$$u_{out} = \frac{kT}{q}\ln\left(\frac{I_c}{I_s}\right) \quad [7.16]$$

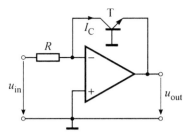

Figure 7.12 Logarithmic amplifier

where k is the Boltzman constant ($1{,}38 \times 10^{-23}$ J/K), T the temperature in Kelvins and q the charge of the electron ($1{,}6 \times 10^{-19}$ As), I_c the collector current of the transistor, and I_s the reverse saturation current of the transistor.

7.3.7 Voltage Follower

The op-amp in the voltage follower application is not used for amplification purposes, but as a buffer amplifier (unity gain) to eliminate loading effects, as when a device with high source impedance is connected to a device with low input impedance (Figure 7.13). Thus, the output voltage is equal to the input voltage $u_{out}=u_{in}$, (the output voltage *follows* the input voltage), but the input impedance is greatly increased (depending on the actual op-amp used, but for more precise applications, this impedance can be greater than 1 TΩ).

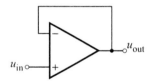

Figure 7.13 Voltage follower

7.3.8 Difference Amplifier

This amplifier should not be confused with the differentiator amplifier described in Chapter 7.3.5.

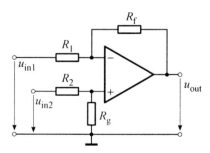

Figure 7.14 Difference amplifier

The difference amplifier is used to subtract two voltages, each multiplied by some value determined by resistances in the circuit (Figure 7.14). The output voltage is equal to:

$$u_{out} = \frac{(R_f + R_1)R_g}{(R_g + R_2)R_1} u_{in2} - \frac{R_f}{R_1} u_{in1} \qquad [7.17]$$

7.3.9 Instrumentation Amplifier

The instrumentation amplifier is a type of difference amplifier with the addition of two input buffers (op-amps on the left) to each input of the difference amplifier. This eliminates the need for input impedance matching. The resistor R_{gain} is used to increase the buffer gain from unity to some desired value using switch selected resistors or potentiometer. As this amplifier has some other favorable characteristics, such as low DC offset, low noise, low drift, very high open-loop gain, very high common-mode rejection ratio, and very high input impedances, it is especially suitable for use in measurement and test equipment where accuracy and stability are required.

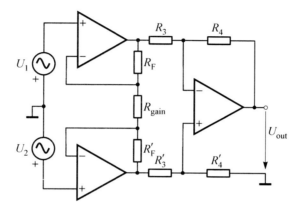

Figure 7.15 Instrumentation amplifier

The output voltage is:

$$U_{out} = (U_2 - U_1)\left(1 + \frac{2R_F}{R_{gain}}\right)\frac{R_4}{R_3} \qquad [7.18]$$

7.3.10 Active Guard

In high-Ohm resistance measurements systems especially, but generally in any application where two conductors at different potentials are not insulated well, a leakage current will flow, causing measurement errors. This problem can be solved if the voltage across the leakage impedance is reduced to zero. The usual method to do this is the application of active guarding (Figure 7.16). The active guard eliminates not only the resistance leakages, but also capacitive leakages. The operational amplifier is used as a voltage follower (unity gain) (Chapter 7.3.7), in such a way that the shield potential follows the guarded conductor potential.

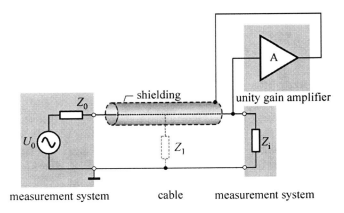

Figure 7.16 Active guarding

In this case, no leakage current will flow through impedance Z_1, and even if the unity gain is not exactly 1, the leakage resistance will be greatly increased, and the leakage current consequently decreased.

7.3.11 Current to Voltage Converter (Transimpedance Amplifier)

In this fairly simple circuit, the input current is converted to the output voltage. As can be seen in Figure 7.17, practically no current flows to the operational amplifier input due to the almost infinite input resistance of op-amp. Also, as the non-inverting input is grounded, the input of the inverting input is held at virtual zero. Therefore, as the whole current flows through resistor R, and as one end of the resistor is at virtual zero, the output voltage is equal to the voltage drop on the resistor:

$$u_{out} = -i_{in} R \qquad [7.19]$$

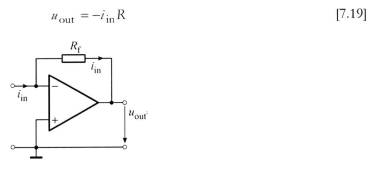

Figure 7.17 Current to voltage converter using op-amps

7.3.12 Voltage to Current Converter (Transconductance Amplifier)

DC current signals are often used in industry to represent some physical quantity. Figure 7.18 shows a simple voltage to current converter in which the voltage input usually represents the voltage output of some transducer or sensor. The voltage range is from 1 V to 5 V, which converts to current from 4 mA to 20 mA (0% - 100% of the measurement range). The 250 Ω resistor establishes the relationship between the current and voltage, so the resistance of the load does not matter as long as the op-amp has a high enough power supply voltage to output the voltage necessary to get 20 mA flowing through it.

AMPLIFIERS IN MEASUREMENT TECHNOLOGY

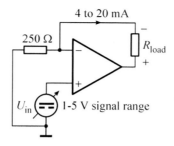

Figure 7.18 Voltage to current converter using op-amps

7.4. MEASURING INSTRUMENTS USING OPERATIONAL AMPLIFIERS

In Chapter 4, the most important analog instruments were described. Such instruments can be enhanced (regarding range, precision, sensitivity, etc.) by using operational amplifiers and other elements such as diodes and thermocouples; several such instruments will be described in this chapter. They can be divided into DC and AC electronic measuring instruments, and most of them can be easily assembled in a basic electrical laboratory.

7.4.1. DC Electronic Voltmeters

DC electronic voltmeters have high sensitivity, high input resistance, and better accuracy than the classic analogue voltmeter. In Figure 8.1 the principle scheme of a simple DC electronic voltmeter is shown; it can be used for voltage measurements from 10 mV to 1 V. With this instrument, currents can also be measured from 10 mA to 1 A as voltage drop on a precision resistor of 1 Ω.

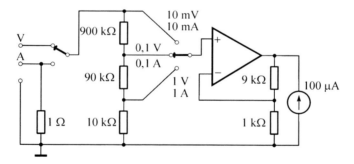

Figure 7.19 Electronic DC voltmeter

The simple DC voltmeter as presented in Figure 7.19 is not suitable for measuring small DC voltages due to the effect of **zero position drift** in DC amplifiers, which can reach up to 1 mV. The zero position drift occurs due to changes in temperature, voltage, and characteristics of embedded elements. This phenomenon cannot be resolved with feedback, and for measuring a small DC voltage, an amplifier with a switch must be used; this switching can be **mechanical**, with **semiconductors** or **photo-resistors**. Figure 7.20 shows the amplifier with a mechanical chopper. The anchor of switch **P** vibrates several hundred times per second between contacts **A** and **B**, and alternatively the input and output of the AC amplifier is short circuited.

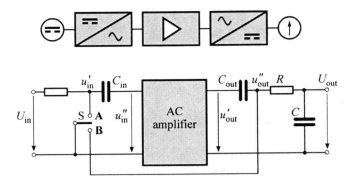

Figure 7.20 Electronic DC voltmeter with chopper

With this chopping, the DC voltage is converted into a pulsating voltage (from zero voltage to input voltage), and after passing through the capacitor C_{in}, only the AC component of the signal remains. The AC voltage is amplified and rectified with the contact action **B**, after which it is then smoothed by the applied RC circuit (Figure 7.21). With this chopping technique, the voltage drift is reduced to only a few microvolts.

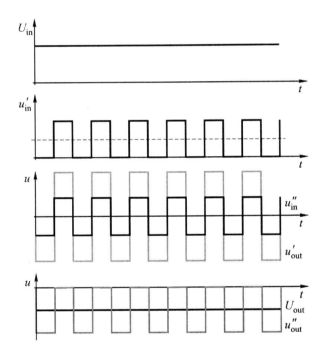

Figure 7.21 Input and output voltages in DC electronic chopper

7.4.2 AC Electronic Voltmeters

AC electronic voltmeters can be divided into two basic types:

    a) **rectifier - amplifier**, where the input voltage is first rectified, mostly with a diode, and then amplified; they are also called **diode voltmeters**.

    b) **amplifier - rectifier**, where the input voltage is first amplified and then rectified.

AMPLIFIERS IN MEASUREMENT TECHNOLOGY

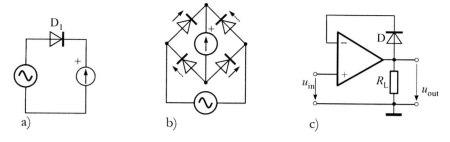

Figure 7.22 Types of rectifiers (from left to right - half-wave rectifier, full-wave rectifier and precision rectifier)

In the rectifier-amplifier type, it is necessary to convert the AC voltage to DC first. The rectified voltage may be proportional to the average, peak, or effective value of the input voltage. There are half-wave and full-wave rectifiers, as well as a precision rectifier (Figure 7.22). The voltage drop V_F across the forward biased diode D_1 in the circuit of a half-wave passive rectifier (7.22a) is undesired. In the precision rectifier version (7.22c), the problem is solved by connecting the diode in the negative feedback loop. The op-amp compares the output voltage across the load with the input voltage and increases its own output voltage with the value of V_F. As a result, the voltage drop V_F is compensated and the circuit behaves almost as an ideal diode with $V_F = 0$ V.

Half-wave rectifying voltmeter with a mean value response is shown in Figure 7.23.

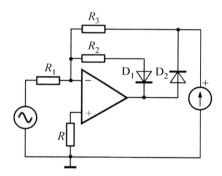

Figure 7.23 Half-wave rectifying AC voltmeter (mean value response)

During the positive half-wave, diode D_1 conducts the current. Since its resistance in the forward direction is small, amplification will be small as well, so the output voltage will be negligible. During the negative half-wave, diode D_2 conducts the current, and in this case the output voltage will be equivalent to the output voltage of inverting amplifiers with resistance R_3 in the feedback:

$$u_{out} = -u_{in} \cdot \frac{R_2}{R_1} \qquad [7.20]$$

The output voltage can be seen in Figure 7.24.

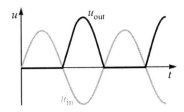

Figure 7.24 Output voltage from the half-wave rectifying AC voltmeter

To achieve **full-wave rectifying**, it is necessary to add one more operational amplifier, which will also conduct the positive half-wave (Figure 7.25).

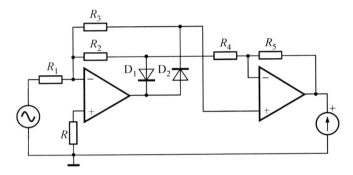

Figure 7.25 Full-wave rectifying AC voltmeter (mean value response)

The output voltage can be seen in Figure 7.26.

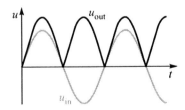

Figure 7.26 Full-wave rectifying electronic AC voltmeter (mean value response)

Full-wave rectifying can be done using the diode bridge or the **Greatz** circuit (Figure 7.27).

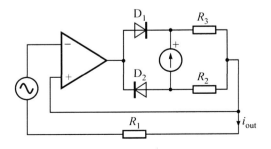

Figure 7.27 Full-wave rectifying electronic AC voltmeter with Greatz circuit (mean value response)

During the positive half-wave, diode D_1 conducts the current and during the negative half-wave, diode D_2 conducts the current. Feedback is achieved by the voltage drop of current $i_{out} = \dfrac{u_{in}}{R_1}$ on the resistor R_1. The output current has a linear relationship with the input voltage, so that such an instrument will have a response to the **mean** value of current; in addition, non-linearity due to diode characteristics is avoided.

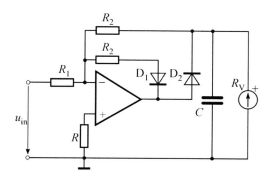

Figure 7.28 Voltmeter with response to the peak value

An example of the **voltmeter with a response to the peak value** using operational amplifiers can be seen in Figure 7.28. There is a capacitor added to the output, which is charged to the peak value of input voltage U_m. If there is a sufficiently large resistance parallel to this capacitor (e.g. input resistance of voltmeter R_V), the voltage on the capacitance will remain proportionate to the value of input voltage U_m. Therefore, it is necessary that the time constant of CR_V is many times larger than the duration of one period of measured voltage. Such voltmeters are calibrated in effective values of sinusoids, and show $\dfrac{U_m}{\sqrt{2}}$ where U_m is the peak value of the measured voltage. If the waveform of the measured signal deviates from sinusoidal, then the peak value can be calculated by multiplying the voltage readings with $\sqrt{2}$.

7.4.3 AC Voltmeters with Response to the Effective Value

Voltmeters with the response to the **mean** or **peak** values of sinusoidal signals will have an error when measuring non-sinusoidal signals. The error depends on how the measured signal differs from the sinusoid. For the measurement of non-sinusoidal waveforms, it is best to use voltmeters that have a response to the **true effective value** (such instruments are marked as **True RMS**). The effective value of AC voltage or current is the value of DC voltage or current that produces the same amount of heat energy in the same period on load. Such instruments have square characteristics, i.e., the output voltage is proportional to the square of input voltage:

$$U_{out} = k \cdot u_{in}^2. \qquad [7.21]$$

So by measuring the DC voltage $\pm U$ or AC voltage with the same effective value, the true RMS voltmeter will get the same reading on the instrument. Several instruments that are responsive to the real effective value have already been mentioned (instruments with moving iron - Chapter 4.3 and electrodynamic instruments - Chapter 4.4).

There are two basic types of electronic instruments that are responsive to effective value:
   a) instruments with thermocouples
   b) converters based on the definition of effective value

Figure 7.29 shows the principle scheme of a simple voltmeter with thermocouples. The output reading of such a thermocouples voltmeter will be proportional to the square of the effective value of the measured quantity, so the scale of this instrument will have square characteristics.

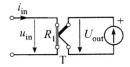

Figure 7.29 Principle circuit of AC voltmeter with thermocouple

To avoid non-linear scale characteristics, two thermocouples are used, as seen in Figure 7.30.

The AC signal is connected to the broadband AC amplifier input A_u, and the heating wire of thermocouples is connected to the amplifier output so that the current through the heating wire is proportional to the measured voltage.

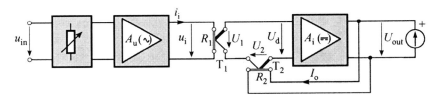

Figure 7.30 AC voltmeter with two thermocouples

In order to eliminate the influence of temperature and non-linearity characteristics of the thermocouples, another thermocouple is used, connected opposite to the first thermocouple. Both of their outputs are connected to the DC amplifier A_i which is heating the wire of the second thermocouple. As thermocouples have the same features, their heating currents and thermal EMF are practically identical. The current through the second thermocouple, which passes through the indicating instrument, is proportional to the effective value of input voltage, so the desired **linear** and not **square** characteristic scale of the instrument is achieved.

The converter that works on the definition of effective values $\left(U_{out} = \sqrt{\frac{1}{T}\int_0^T u_{in}^2 dt} \right)$ is shown in Figure 7.31. This converter has multiplier integrating amplifiers, a square roots circuit, and a low pass filter. The low pass filter can also be made using op-amps. Today, this type of AC voltmeter can be found in integrated technology.

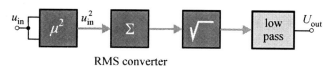

Figure 7.31 AC voltmeter based on effective value definition

Selected bibliography:
1. D. Vujević, B.Ferković, "*The Basics of the Electrotechnical Measurements, part I*", Zagreb 1994
2. *Operational Amplifier Applications*, http://en.wikipedia.org/wiki/Operational_amplifier_applications, Wikipedia, the Free Encyclopedia, Retrieved September 20th, 2010
3. Serhan Yamacli, *Circuits with Resistive Feedback*, http://www.scribd.com/doc/7094428/Positive-Feedback-Opamp, Retrieved September 20th, 2010
4. Roman Honig, "*Practical Aspects of High Resistance Measurements*", CAL LAB, The International Journal of Metrology, Jan-Feb-Mar 2010
5. Klaas B. Klaassen, "*Electronic Measurement and Instrumentation*", Press Syndicate of the Universty of Cambridge, 1996
6. *Voltage-to-Current Signal Conversion*, http://www.allaboutcircuits.com/vol_3/chpt_8/7.html, Retrieved September 20th, 2010
7. Bruce Carter and Thomas R. Brown, *Handbook of Operational Amplifier Applications*, Application Report, SBOA092A, Texas Instruments, 2001
8. *AC Theory*, http://www.opamp-electronics.com/tutorials/ac_theory.htm, Retrieved September 20th, 2010

8. OSCILLOSCOPES

Today oscilloscopes are one of the most-used types of measurement instruments in the world. They can present current values of the signal in two dimensions, usually as a function of time, but also as a function of some other signal (X-Y). Oscilloscopes can measure voltages of different durations, amplitudes, and shapes in wide frequency ranges. Other quantities can be also measured by oscilloscopes after they are transformed to a voltage using some type of signal transducer. The vertical deflection is controlled by the voltage applied to the inputs, while the horizontal deflection is controlled by the sawtooth voltage generator that produces a linear voltage rise as a function of time. This generator allows us to see the voltage as a function of time. Figure 8.1 shows the block diagram of the Cathode Ray Oscilloscope. The basic parts are the systems for horizontal and vertical deflection, the power sources, and the cathode ray tube.

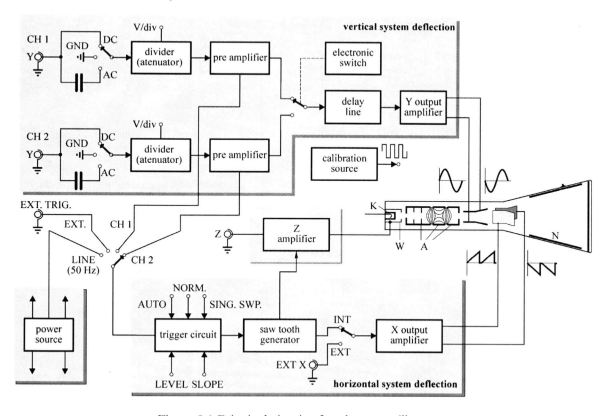

Figure 8.1 Principal circuit of analogue oscilloscope

8.1. CATHODE RAY TUBE

The most important part of the oscilloscope is the cathode ray tube. It is a vacuum glass balloon filled with several electrodes (Fig. 8.2).

Moving from left to right, the cathode is placed at the left end of the tube. The cathode is heated indirectly and dissipates a cloud of electrons. The Wehnelt cylinder is placed around the cathode with a small hole at the tip of the cylinder. Its potential is negative relative to the cathode, causing the electrons to deflect from the cathode and gather in the middle of the cylinder. The potential of the Wehnelt cylinder can be easily changed and used to regulate the number of electrons that exit the cylinder.

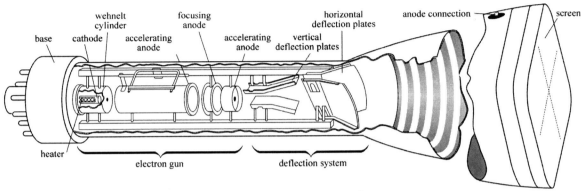

Figure 8.2 Cathode ray tube

This effect is observed as a brightness change of the ray on the screen. The regulation of the Wehnelt cylinder voltage is taken care of by the **INTENSITY** control on the front panel of the oscilloscope. The sufficiently large negative potential with respect to the cathode can completely block the passage of electrons, thus causing the disappearance of the bright spot on the screen.

Following the Wehnelt cylinder in the cathode ray tube is a set of anodes, which are at a high positive potential (several hundred volts), thus forcing the electrons to move towards the display circuit. Usually there are three anodes – two acceleration anodes that are at a higher potential than the third focusing anode, which is located in the middle. Therefore, these anodes create an electric field between them that acts as a lens with a focus on the screen (Fig. 8.3).

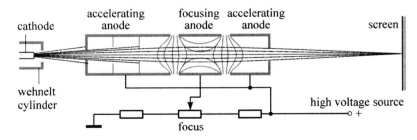

Figure 8.3 Electron gun and focusing anodes

The middle anode is called the focusing anode, and with the change of its potential, it can focus an image on the screen (on the oscilloscope, the slider which controls the focusing anode voltage is

OSCILLOSCOPES

called **FOCUS**). Between the anode and the screen, there are two pairs of deflection plates. Plate couples are set perpendicular to each other, one pair of plates deflecting the electron beam in the horizontal direction and the other in the vertical direction.

The measured voltage is connected to the plates attached to the vertical deflection system, while the sawtooth voltage is connected to the horizontal deflection plates. The electric field acts on the electron beam in the direction of the plates with a positive potential (Fig. 8.4).

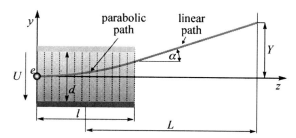

Figure 8.4 Electron beam path

At the same time, the electron beam is affected by the anodes that force the beam in the direction of the screen, causing the electron beam to move in a parabolic curve while inside the plates; after leaving the plates, the beam will continue the linear motion. After the beam leaves the plates, it needs to accelerate to achieve the necessary kinetic energy. For this purpose, as a conductive coating on the walls of the helicoidal circuit, there is an anode for the additional acceleration with voltage of a few kilovolts. The cathode ray tube screen is made of fluorescent material, and it will radiate light after the electron beam impact. The property of the screen causing the screen to remain bright for some time after the impact is called persistence. In recent years, screens with memory that can hold the beam light almost indefinitely have been developed. In front of the fluorescent screen, there is a transparent plate with square mesh, with which the oscilloscope can be used for measurement.

8.2. SYSTEM FOR VERTICAL DEFLECTION

The vertical deflection system has the following main parts:

- input divider
- preamplifier
- delay line
- electronic switch

The input divider is used to adapt the input voltage to a level that corresponds to the working point of the preamplifier. The divider must have independent input and output voltage ratios, and thus a frequency compensated divider is used, whose principle circuit is shown in Figure 8.5. At the input of the divider, there is a switch by which direct input (DC) can be chosen, input through the capacitor (AC) for measuring the AC signal only, or ground (GND). The measured signal is usually connected to oscilloscope with the probe via a coaxial cable.

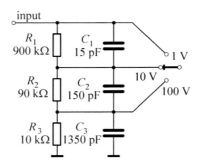

Figure 8.5 Oscilloscope input divider

Passive probes that do not contain any active element are most frequently used. Probes can also reduce the voltage signal and then usually have the tags x10 (to reduce 10 times), or x100 (to reduce 100 times), etc. If the probe does not reduce the input signal, it will have the tag x1. Oscilloscopes for frequencies below 500 MHz usually have an internal resistance R_0 of 1 MΩ in parallel with a capacity C_0 from 8 pF to 50 pF (Fig. 8.6).

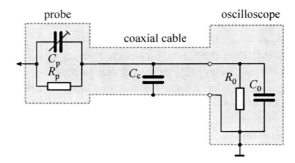

Figure 8.6 Connection of signal to oscilloscope via probe

Probes that reduce the signal 10 times must have the resistance R_P from 9 MΩ in parallel with a variable capacitor C_P. A variable trimmer capacitor is used to compensate the probe, i.e., to achieve the frequency independent ratio:

$$R_P C_P = R_0 (C_0 + C_k). \qquad [8.1]$$

where C_k is the capacity of the coaxial cable. Compensation of the probe (Figure 8.7) is achieved using a rectangular signal of 1 kHz, which is embedded in each oscilloscope. When changing the trimmer capacitors to achieve a rectangular signal on the screen without distortion, the probe is properly compensated, which is shown in Figure 8.7b, while the probe is overcompensated in Figure 8.7a and undercompensated in Figure 8.7c.

Figure 8.7 Compensation of the oscilloscope probe

OSCILLOSCOPES

The preamplifier amplifies the voltage in the frequency range from DC to some upper frequency, to the level necessary to achieve sufficient deflection of the beam on the screen. The amplifier has a balanced output, allowing two voltages of equal amplitude, but mutually phase shifted by 180° to reach the plates, causing the potential in the middle of the plate to always remain zero.

Passing through circuits causes delays of measured and other signals. Signals in horizontal and vertical systems may have different time delays, and it may happen that the beginning of the measured signal is not displayed if the delay in the system for horizontal deflection is larger than the vertical. This is solved by the delay-line which is inserted between the amplifier and preamplifier, which will ensure that the delay in the vertical deflection system is greater than in the horizontal one. The delay line consists of symmetric LC elements (Figure 8.8) joined as "T" elements. Voltage u_0 is a faithful reproduction of voltage u_{in}, but delayed:

$$t_K = \sqrt{LC} \qquad [8.2]$$

Time delays can be chosen by the appropriate number of LC elements.

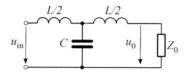

Figure 8.8 Oscilloscope delay line

Oscilloscopes can also be used to observe two signals simultaneously. This can be achieved with an oscilloscope with two beams, but the present cathode tube must have twice as many electrodes. A more frequent and cheaper, but also satisfactory, solution is a single beam oscilloscope with an electronic switch. The electronic switch can operate in the synchronous or asynchronous mode. In the synchronous mode during one full period of sawtooth voltage, only the signal from one input is shown, while in the second period, it is the signal from another input that is shown (Fig. 8.9a).

The synchronous mode is used for high frequency signals, while at frequencies lower than 1 kHz, the asynchronous mode electronic switch is used. In this mode, the electronic switch is controlled by pulses with the frequency of about several tens of kilohertz, so that during one period of sawtooth voltage, the display shows both signals that are assembled from small segments (Figure 8.9b).

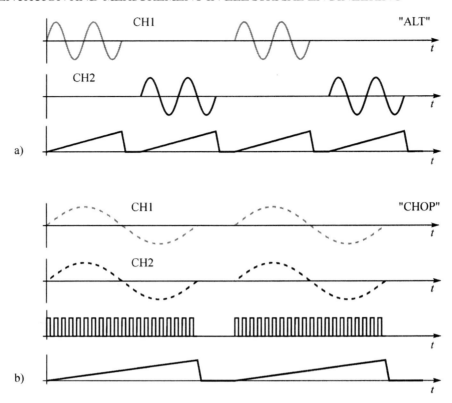

Figure 8.9 Synchronous and asynchronous oscilloscope switching modes

8.3. SYSTEM FOR HORIZONTAL DEFLECTION

The horizontal deflection system consists of:

- trigger circuit
- sawtooth voltage generator
- horizontal amplifier

If the desired signal should be observed as a function of time, the measured signal must be brought to the vertical plate pair of plates and the sawtooth voltage must be brought to the horizontal plates. The sawtooth voltage will rise and provide uniform speed motion of signal from left to right. When the signal reaches the right end, it must return to the left end as fast as possible (Figure 8.10).

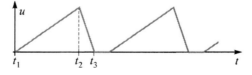

Figure 8.10 Sawtooth voltage

The sawtooth voltage is generated by charging the capacitor with constant current, and then quickly discharging it (Figure 8.11).

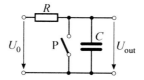

Figure 8.11 Sawtooth generator

To obtain a still image on the screen, the ratio between the frequency of the measured signal and the frequency of the sawtooth signal must be an integer. Such a relationship is called synchronization. To achieve synchronization, the measured signal must have some influence on the sawtooth voltage. This is achieved by a trigger circuit, which is controlled by the measured signal, resulting in a strong relationship between the measured signal and the sawtooth voltage.

The trigger circuit, along with the measuring signal, can also be controlled with other signals connected to the oscilloscope, or with a power supply. The oscilloscope must also enable the observation of short transient periods by triggering only one period of the sawtooth voltage. The point on the measured signal to set the trigger is determined with two buttons: slope and level. Only the level is not sufficient to unambiguously determine the point of periodic signals (e.g., the sine-wave signal in one period has the same level in two different points, but a different slope, and this is valid for any periodic signal) (Figure 8.12).

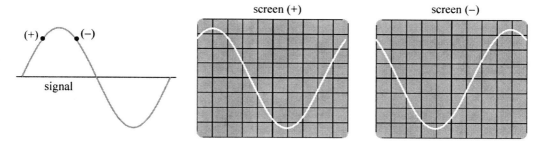

Figure 8.12 Display depending on slope choice

The horizontal amplifier has a similar function as the amplifier for vertical deflection. The deflection ratio is constant and usually has two stages x1 and x10. With the help of this, we can stretch the measured signal in x - direction.

8.4. DESCRIPTION FRONT Oscilloscope Panel (TEKTRONIX 2205)

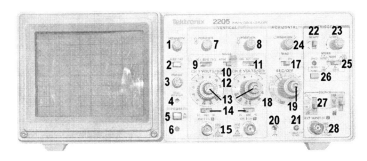

Figure 8.13 Oscilloscope Tektronix 2205 from the Faculty of Electrical Engineering and Computing

Oscilloscope Tektronix 2205 (Figure 8.13) is a common type of oscilloscope with a cut-off frequency of 20 MHz with two inputs.

Table 9.1. Description of the Tektronix 2205 front panel

Number	Mark	Description
1	INTENSITY	Adjusts the brightness of the electron beam on the screen (connected to the slider, which changes the Wehnelt cylinder potential)
2	BEAM FIND	Finds spots on the screen
3	FOCUS	Focuses the electron beam (connected to the slider, which changes the potential of the focusing anode)
4	TRACE ROTATION	Adjusts the direction of the image tilt on the screen using the screwdriver
5	POWER ON/OFF	Switches to turn on/off the oscilloscope
6	VERTICAL ⇕ POSITION	Shifts the position of spots in the y direction separately for each of the two inputs by applying DC voltage to the plates
7	CH1 BOTH CH2	Chooses the input that will be shown on the display oscilloscope CH1 input 1 BOTH - both inputs CH2 - input 2
8	NORM INVERT CH2	Viewing another channel signal on the screen NORM - normal INVERT - inverted
9	ADD ALT CHOP	Operates the electronic switch ADD - The display shows the sum of the voltage of both inputs. The display shows the sum of two signals CH1 + CH2 if the position of switch under number 8 is set to NORM, or the difference of signals CH1-CH2 if the switch under the number 8 is set to position INVERT.

Number	Mark	Description
		ALT - synchronous mode switch
		CHOP switch - asynchronous mode switch
10	CH1 VOLTS/DIV CH2 VOLTS/DIV	Adjusts the ratio between input voltage and actual deflection of the signal on the screen. Using this data and by measuring the signal on the screen, it is possible to determine the voltage measured at the inputs of channel 1 or 2.
11	VAR	VAR potentiometer is located inside the switch VOLTS / DIV and is used to continuously adjust the degree of each factor ratio. During the measurement, the potentiometer must be in position CAL (calibrated).
12	AC GND DC	Chooses the type of input DC – direct input AC line input - input through capacitors to show only the AC signal GND - grounded input is used to adjust the reference level of signal
13	HORIZONTAL ⇐ POSITION ⇒	Shifts the position of signal in the direction of the x-axis separately for each of the two inputs by applying DC voltage to the horizontal plates
14	MAG x1 x10	Chooses the horizontal deflection amplification factor
15	SEC/DIV	Adjusts the duration of sawtooth voltage rise, by which the frequency of the observed signal can be measured. In the x-y position, the signal is viewed as a function of another signal, and not as a function of time.
16	PROBE ADJUST ≈ 500 mV Peak to Peak 1 kHz	Rectangular output voltage to adjust (compensate) measurement probe
17	⊥	Connection to grounding
18	SLOPE _/‾ ‾_	Switch to select the slope at the trigger level _/‾ positive slope ‾_ negative slope
19	LEVEL	Potentiometer to adjust the trigger level
20	TRIGGER AUTO NORM SGL SWP	Trigger mode AUTO mode switch - Constant work of sawtooth generator voltage without the presence of measured voltage. Used to adjust the initial position without the presence of measured signal NORM – The work of the sawtooth generator using the trigger circuit. SGL SWP - Triggering only one period of sawtooth voltage to observe transients. Triggering is done each time by pressing the RESET button

Number	Mark	Description
21	SOURCE	Selection of sources for triggering
	CH1	CH1 - input 1
	CH2	CH2 - input 2
	EXT	EXT- external source that can be power line (symbol LINE) or voltage connected to the input marked with the EXT INPUT

8.5. MEASUREMENT USING OSCILLOSCOPES

Using the oscilloscope, several quantities can be measured:

- **Voltage** (AC and DC) measurement is done by multiplying the factors of factor ratio (VOLTS/DIV) with vertical spacings on the screen.
- **Current** is measured using current probes or by measuring the voltage drop on the precise resistor
- **Time intervals** (e.g. measuring the width of a rectangular pulse) are measured by multiplying the sawtooth voltage rise time (SEC/DIV) with the number of horizontal spacings
- **Frequency** of a signal is determined by measuring the periods of a signal using the formula:

$$f = \frac{1}{T} \qquad [8.3]$$

where T is the period of a signal.

- The **phase angle** between two signals is determined by measuring the signal period T and delay time t_K. (Figure 8.14.); the phase angle φ is then determined using the expression:

$$\varphi = \frac{t_K}{T} \cdot 360° \qquad [8.4]$$

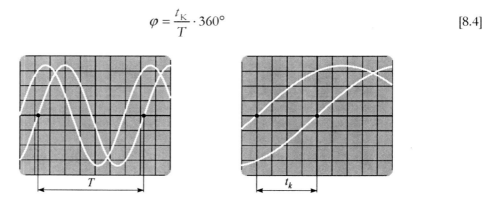

Figure 8.14 Phase angle measurements with oscilloscope

8.6 DIGITAL STORAGE OSCILLOSCOPES (DSO)

The input circuit of the digital oscilloscope is more or less the same as in the analogue oscilloscope. The sampling circuit tracks the input signal and switches into the hold state in a well-defined moment (track&hold). The signal is then sampled with the A/D converter to digital values, which are stored in

OSCILLOSCOPES

memory for analysis and display. The horizontal system's clock determines the sample rate of the A/D converter. The sampled values, as well as the numerically interpolated intermediate values, are displayed in either the cathode ray tube or on the LCD display (Figure 8.15). Digital oscilloscopes can have many additional functions not found in analogue oscilloscopes, especially for data processing and enhancing the display quality.

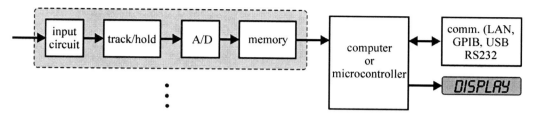

Figure 8.15 Digital storage oscilloscope block diagram

8.6.1 Sampling Methods

The use of an appropriate sampling method is important for accurate reconstruction of the input signal. The first imperative is that the sampling theorem is fulfilled, which means that the sampling frequency must be at least twice as high as the highest harmonic part of the measured signal; this is often not fulfilled. The sampling frequency is determined as:

$$f_S = \frac{N_{MEM}}{T_{REC}} [\text{samples/second}] \qquad [8.5]$$

where f_S is the maximum sampling frequency, N_{MEM} is the number of samples that fits in the memory, and T_{REC} is the record time length. Today, oscilloscopes with a sampling frequency of several gigasamples per second are common.

If the sampling frequency is not high enough, the aliasing error occurs, which distorts the input signal (Figure 8.16). The signal displayed has lower frequency then the input signal.

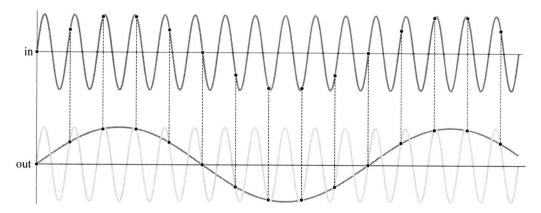

Figure 8.16 Aliasing error

The oscilloscope can easily display slowly changing signals, as it will collect enough sample points in real time to make a good representation of the input signal. With the same sampling frequency it will collect smaller and smaller number of samples. Depending on the maximum sampling rate, there will be a maximum frequency that the oscilloscope can measure with satisfactory accuracy. This type of sampling is called **real time sampling with interpolation**, as the points between the samples are determined by the interpolation technique (Figure 8.17).

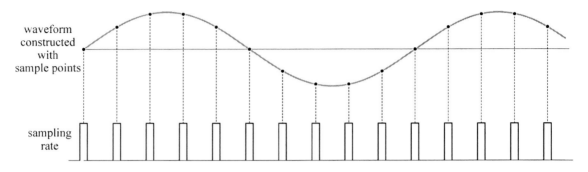

Figure 8.17 Real time sampling with interpolation

This is the only possible method for single-shot or transient signals. However, for a high frequency signal that repeats itself, the oscilloscope cannot use this method due to aliasing errors. It therefore uses another sampling method called **equivalent-time sampling method,** in which it builds the signal over time by taking one or several samples in each repetition, but at a different position (Figure 8.18).

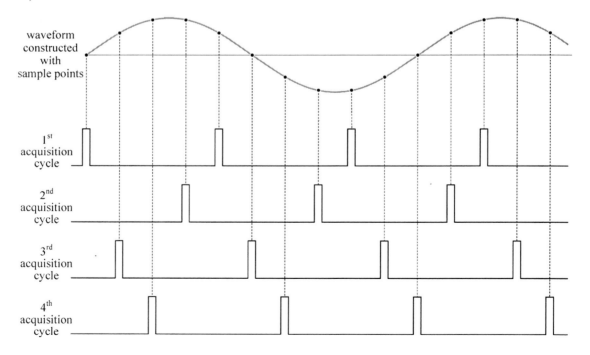

Figure 8.18 Equivalent time sampling method

Selected bibliograpy:
1. D. Vujević, B. Ferković, *"The Basics of the Electrotechnical Measurements"*, part I, Zagreb 1994
2. V. Mamula, *"Measurements in Electrical Engineering"*, (in Croatian), Split, 1986
3. D. Vujević, *"Measurements in Electrical Engineering – Laboratory Experiments"*, (in Croatian), Zagreb, 2001
4. Tektronix Oscilloscope 2205, *User manual*, Tektronix, 1990
5. *How Oscilloscopes Work*, Oscilloscope Guide, BK Precision, http://www.bkprecision.com/download/scope/HowScopeWork.pdf, Retrieved October 10th, 2010
6. *Oscilloscope Calibration*, Fluke Application Note, http://www.testforce.com/testforce_files/app_notes/Oscilloscope_Calibration.pdf, Retrieved October 10th, 2010
7. *Digital Oscilloscopes*, http://www.electro-tech-online.com/electronic-theory/56-oscilloscope.html, Retrieved October 10th, 2010
8. Carl von Ossietzky, *Cathode Ray Oscilloscope, Digital Storage Oscilloscope and Function Generator*, Universität Oldenburg – Faculty V - Institute of Physics Module Introductory Laboratory Course Physics – Part I, http://physikpraktika.uni-oldenburg.de/download/GPR/pdf/E_Oszilloskop.pdf, Retrieved October 10th, 2010
9. *Analogue and Digital Oscilloscope*, Czech Technical University, Department of Measurement, http://measure.feld.cvut.cz/en/system/files/files/en/education/courses/AE0B38DCZ/AE0B38DCZ_L3.pdf, Retrieved October 10th, 2010

9. DIGITAL INSTRUMENTS

Unlike analogue measuring instruments that indicate measuring results by using a pointing device and scale, digital measuring instruments show a measurement result by using numerical digits. The numerical result is appropriate if one wishes to express the measurement result in multiple digits (e.g. 5 or 6 digits); such resolution with analogue instruments is generally not possible. The digital display of measurement data in analogue instruments appears in some form (e.g., showing consumption of electricity in induction electricity meters or the markings on the precise resistance decades). Today, digital measuring instruments have entered most areas of measurement applications. The advantages of digital measuring instruments are numerous:

Figure 9.1 Precision voltmeter Agilent 3458A with 8 ½ digit resolution at the Faculty of electrical engineering and computing in Zagreb

- the resolution and accuracy of digital instruments is better than the resolution and accuracy of analogue instruments (up to 8 1/2 digits in the best digital voltmeters)
- possibility of a larger number of readings during a time interval - some voltmeters have 100,000 readings per second
- possibility of automation of test procedures and automated processing of measurement data
- a large input impedance.

Digital instruments also have some drawbacks. For example, they need a stable power supply, and it is relatively difficult to read more than one digital measuring instrument at the same time.

An important feature of digital instruments is the **number of digits** that can be displayed on the display. The number of digits is generally expressed as X 1/2 digits, where 1/2 is a partial digit. This means that, e.g., the instrument of 3 1/2 digits in the measuring range of 300 mV can display from 000,0 mV to 299.9 mV, because the first digit cannot display digits "0" through "9", but only the numbers "0","1", and "2". Today, the most advanced digital multimeters have 8,5 digits resolution (Figure 9.1).

Due to construction reasons, digital measuring instruments use a binary counting system, with 2 as its base, in contrast to the common decade numbering system. The numbers are then expressed as the sum of powers of the number 2. The binary system requires only two digits, which can easily be achieved by using elements that can have two states (e.g. switches, transistors, relays, etc.). Measuring results are then transformed to decimal numbers before they are displayed, as decade numbers are more suitable for reading by the operator. A block diagram of a digital multimeter is shown in Figure 9.2, while the measurement unit is shown in Figure 9.3.

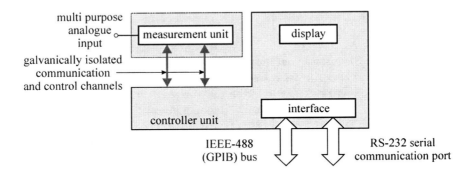

Figure 9.2 Block diagram of a digital multimeter

Digital measuring instruments transform the analogue signal to digital, which is then displayed on the numeric display of the measuring instrument. This function is performed by **analogue-digital converters** (A/D converters), which are the most important part of a digital measuring instrument. Among all the other quantities, the most appropriate for the conversion is the DC voltage; all other quantities are transformed to DC voltage before conversion takes place.

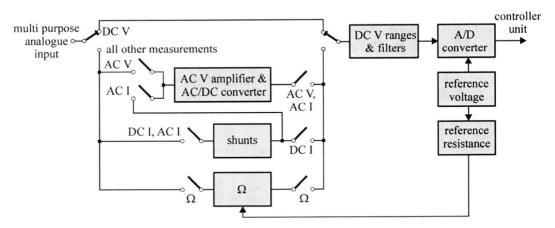

Figure 9.3 Measurement unit of the digital multimeter

High-resolution digital voltmeters are widely used in measurements, and input parameters can contribute to measurement results. Therefore, to make measurement uncertainty as small as possible, it is necessary to characterize its input parameters. The parameters affecting the ideal input of DVM input can be seen in Figure 9.4.

DIGITAL INSTRUMENTS

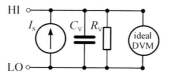

Figure 9.4 Input circuit of DVM (Input resistance, input capacitance, and input offset current)

The different effects in the DVM input can be characterized as input resistance, input capacitance, and input offset current. For the best DVM, the input offset current should be in the range of picoamperes, the input resistance in the range of teraohms, and the input capacitance in the range of a few hundred picofarads. As the main quantity in the digital multimeter to be quantized is DC volts, all other quantities must somehow be converted to DC voltage. The current I is measured as the voltage drop U_S on precision resistor R_S using the non-inverting operational amplifier circuit, as seen in Figure 9.5.

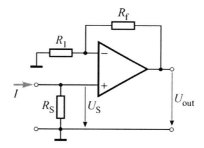

Figure 9.5 Current measurements in DVM

The resistance is measured according to Figure 9.6. The measurement can be performed with two wire connections for larger value resistors, and also with four wire connections for low-value resistance measurements.

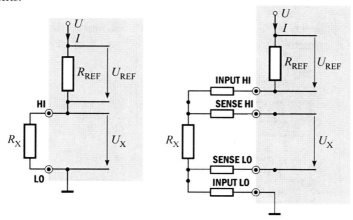

Figure 9.6 Resistance measurements in DVM

The unknown resistance is compared with the known resistance R_F. The ranging for current and resistance measurements can be performed using the different values of shunts and resistors.

9.1 ANALOGUE TO DIGITAL CONVERTERS

Analogue to digital converters (A/D) base their operation on many different principles. Several major types of A/D converters will be discussed in the following chapters.

9.1.1 A/D Converter of Voltage to Time

In the A/D converter of voltage to time, the time sawtooth voltage U_P (or some other voltage that rises linearly with time) needed to reach the measured voltage U_X (Figure 9.7) is measured. The sawtooth voltage starts to rise from some negative value and at time t_1 reaches a value equal to zero, and at time t_2 it reaches the value of U_X. The measured voltage U_X is then proportional to the time interval $\Delta x = t_2 - t_1$. Time intervals can be very accurately measured by electronic counters. The counter is connected to the high-frequency oscillator via an electronic switch (Figure 9.8).

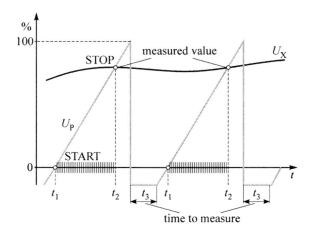

Figure 9.7 Conversion of voltage to time

At time t_1 the null-indicator $\mathbf{N_1}$ closes the electronic switch and the timer starts counting pulses from the oscillator. The counting will stop when the null-indicator $\mathbf{N_2}$ at time t_2 opens the electronic switch. The number of pulses counted by the counter is then proportional to the measured voltage U_x. This converter measures the **current** value of the measured voltage. A block diagram is shown in Figure 9.8.

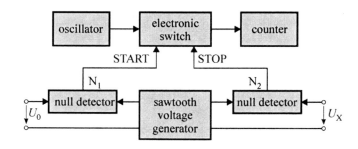

Figure 9.8 Block diagram of A/D converter of voltage to time

DIGITAL INSTRUMENTS

9.1.2 Dual Slope (Integrating) A/D Converter

The dual slope A/D converter has great accuracy and low sensitivity to noise, and is often used in digital voltmeters. The block diagram is shown in Figure 9.6, while the principle of work can be seen in Figure 9.7.

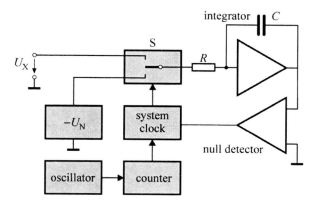

Figure 9.6 Dual slope A/D converter block diagram

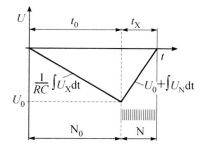

Figure 9.7 Dual slope A/D converter's principle of work

The measured voltage in the integrating amplifier is integrated during the predefined time t_0, and the voltage at the output is:

$$U_0 = \frac{1}{RC} U_X t_0. \qquad [9.1]$$

After that, the switch **S** switches to the precisely known voltage $-U_N$. U_N voltage is then integrated during the time t_X until the voltage at the output falls to **zero**. Because of constant voltage, the U_N capacitor discharge slope is always constant and does not depend on the measured voltage U_X. The counter counts N_0 impulses during the time period t_0, and during the time period t_X, it counts N pulses. Dual slope A/D converter measures the **average value** which is:

$$U_X = U_N \frac{N}{N_0}. \qquad [9.2]$$

As it is clear from [9.2], the measured voltage does not depend on the **RC** time constant, nor the accuracy of the oscillator, but only on the accuracy of the precise voltage U_N and counting of N_0 and N impulses, which is not a major problem.

9.1.3 Successive Approximation A/D Converter

Successive approximation A/D converters consist of a group of precision resistors, a sensitive null-detector, and a control device which performs the balancing (by connecting and disconnecting the resistances until equilibrium is reached). Switching is performed by an electronic switch, and resistor grading can be binary or decimal.

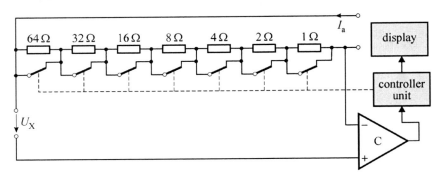

Figure 9.8 Successive approximation A/D converter

The result is read from the switch positions after balancing. The principle scheme of the converter is shown in Figure 9.8, and the principle of balancing in Figure 9.9.

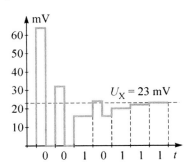

Figure 9.9 Principle of work of successive approximation A/D converter

For example, how do we measure voltage $U_X=23$ mV with the A/D from figure 9.8?
The converter has seven binary scaled resistances from 1 Ω to 64 Ω and the auxiliary current is $I_a=1$ mA. At the beginning, all of the switches are opened, which corresponds to a state of "0". The comparison starts from the most important digit.

When the switch parallel to resistance of 64 Ω is closed, the compensation voltage will be $U_k = 64$ mV and larger than the measured voltage (64 mV > 23 mV). The switch is turned back to the previous position (open), and the switch parallel to resistance of 32 Ω is closed. Since the voltage at this resistance is again higher than the measured voltage (32 mV > 23 mV), the procedure is repeated. The switch opens and closes the switch parallel to resistance of 16 Ω. Since the output voltage is now smaller than the measured voltage (16 mV < 23 mV), the switch remains closed, which corresponds to a state of "1", and the converter closes the next switch parallel to resistance of 8 Ω. Now the total voltage is again higher than the measured voltage and the switch returns to the previous state (16 mV + 8 mV = 24 mV > 23 mV). The procedure is repeated until the last of the resistance switches changes state or until the null-detector indicates that balancing is done. In this case, this is achieved when the switches parallel to resistances 64 Ω, 32 Ω and 8 Ω stay open (state "0"), and switches parallel to all other resistances are closed (state "1"), which will be equaled to the measured voltage: 16 mV + 4 mV + 2 mV + 1 mV = 23 mV or $23_{10} = 0010111_2$. The successive approximation converter measures the **current** value of voltage. The actual A/D converters have binary graduated voltages obtained from a digital analogue converter instead of binary graduated resistances as seen in Figure 9.10.

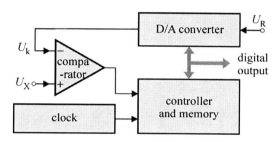

Figure 9.10 Successive approximation A/D converter using D/A converters instead of resistor banks

9.1.4. Parallel A/D Converter

The A/D converter is used when it is necessary to convert the analogue signal to digital form as fast as possible (e.g., digital oscilloscopes). The A/D converter consists of a series of parallelly connected comparators. The **comparator** (Figure 9.10) is an operational amplifier without feedback, whose output will be logic "1" or + supply voltage if the voltage at the non-inverting terminal is higher than the voltage at the inverting terminal; otherwise, the output will be logic "0" or - supply voltage (Figure 9.11), so the comparator essentially compares two voltages and indicates which one is higher.

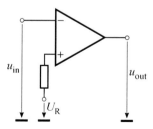

Figure 9.10 Comparator

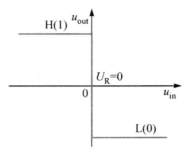

Figure 9.11 Comparator input and output voltage

A converter with **n** bits has 2^n-1 comparators, so e.g., the 4-bit converter has 15 comparators. Figure 9.12 shows the principle scheme of a 3-bit parallel A/D converter that has seven comparators. Reference voltage $U_N=8$ V is divided by precision resistor divider into 7 levels, from 1 V to 7 V. In comparators where the measured voltage U_X is higher than the reference level voltage, the output will have state "1," and other outputs will remain in state "0." The outputs of comparators are connected to the encoder that gives results in binary form. For example, for the voltage of $U_X = 4.5$ V, the comparator outputs of K_1, K_2, K_3 i K_4 will be "1" and the others "0."

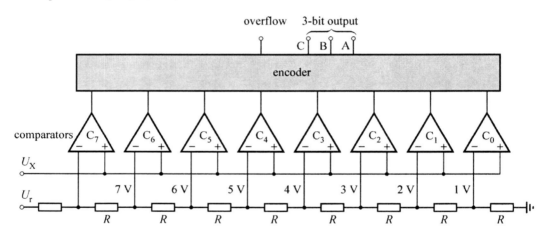

Figure 9.12 Parallel A/D converter

9.2 AC MEASUREMENT IN DIGITAL MULTIMETERS

There have been many different techniques for the measurement of AC voltages in multimeters. As already discussed, there is a difference in the mean value and RMS value of an AC signal. The RMS value is more useful than the mean value, so it is important to know how a particular multimeter makes the AC measurements. A True RMS (root-mean-square) AC measurement is often described as a measure of the equivalent heating value of the signal, with a relationship to the power dissipated by a resistive load driven by the equivalent DC value. This "heating" potential has been measured in different ways over the years:

- Applying the AC signal to a thermopile and measuring the resultant temperature. This technique has a low speed of conversion, and does not have good characteristics at low frequencies.

DIGITAL INSTRUMENTS

- Using hardware to generate the root, mean, and square functions in an AC-to-DC converter. This technique is fast, accurate, and relatively inexpensive to implement, and the converter can fit on a single chip. However, there are some limitations concerning bandwidth, crest factor, etc. This method is therefore not appropriate for low frequency measurements (Figure 9.13).

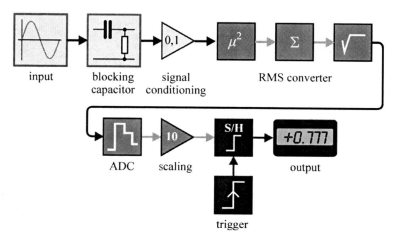

Figure 9.13 Analogue AC converter

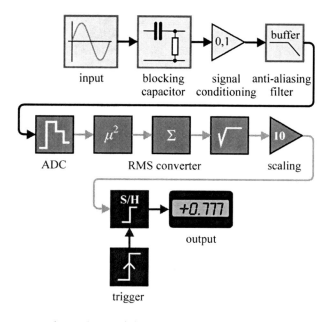

Figure 9.14 Digital sampling AC converter

- Using digital sampling technique to measure AC (Figure 9.14). This technique is the most popular today, as it provides very good linearity, high accuracy, and is applicable to all types of AC signals. The technique is similar as in digital oscilloscopes (Chapter 8.6.1). An anti-aliasing filter is placed after signal conditioning, and the output signal is digitized at a high sample rate with the A/D converter. The RMS value is then calculated from the sample values. The digital sampling technique is generally faster than the analogue converter technique. However, the technique deteriorates at high frequencies because of the anti-aliasing filter, and harmonics above the bandwidth of the filter will be lost.

Some of the analogue techniques have been already mentioned earlier in this book (Figures 7.29, 7.30 and 7.31).

9.3 MAIN CHARACTERISTICS OF DIGITAL INSTRUMENTS

Features of digital instruments vary from instrument to instrument, so for each specific instrument, an instruction manual must be consulted. Error limits of digital instruments are not determined by the accuracy class as for analogue instruments, but are stated by:

- resolution
- constant error
- proportional error

Resolution is the number of digits shown on the display. The constant error remains the same for the entire measurement range of the instrument. It can be expressed as the number of digits and/or percentage of the measurement range. The proportional is usually given as a percentage of the measured value. The error of each digital instrument can be expressed with constant error, proportional error, or a combination of both.

Example 10.1. A digital voltmeter of 3 ½ digits has error limits ± (1% of reading + 10 digits) on the measuring range of 120 V. What are the error limits, if the measured voltage is 10 V?

Error limits proportional to the measured value (1% of reading) are easily calculated (1% of 10 V = 0.1 V). Error limits of errors given in digits must be converted to units measured $\left(\dfrac{10}{1200} \cdot 120\,\text{V} = 1\,\text{V}\right)$, where 1200 is the total number of digits in a given measurement range, and 120 is the measurement range in measured quantities. The total error limits are, therefore, 0.1 V + 1 V = 1.1 V.

The operating manual of each instrument contains other important specifications of digital instruments. Particularly important data are the **input impedance** of the digital voltmeter, which is usually in the range from 10 MΩ to over 10 GΩ. Other useful data include the number of readings per second, which can reach up to 100,000 readings per second, noise suppression, and connection possibilities to computers.

9.4 ELECTRONIC WATTMETER

Electronic wattmeters consist of a transducer that converts the measured power to the standardized DC current or voltage, which can be measured with digital voltmeter or acquired by the analogue to digital converter (Figure 9.15). It multiplies the voltage and current in real time to obtain the power. These instruments are characterized by low consumption and narrow limits of error.

DIGITAL INSTRUMENTS

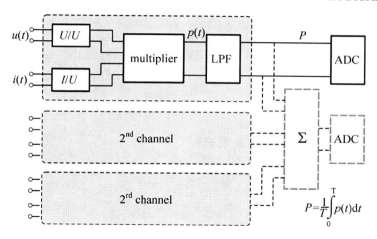

Figure 9.15 Electronic wattmeter

The electronic wattmeter uses several different kinds of multipliers, most often the **time division multiplier** (Figure 9.16).

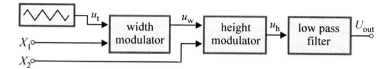

Figure 9.16 Time division multiplier principle circuit

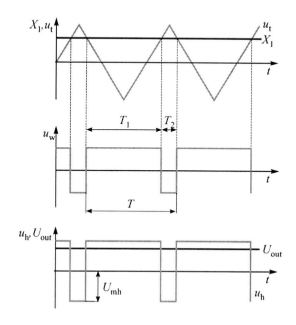

Figure 9.17 Time division multiplier principle of work

131

The time division multipliers produce rectangular pulses whose amplitude is proportional to current, and whose width is proportional to voltage. The area of pulses is proportional to power. By smoothing the impulses with low pass filters, a voltage that is proportional to the measured power is obtained (Figure 9.17). The output voltage is equal to:

$$U_{out} = k \cdot X_1 \cdot X_2 \qquad [9.3]$$

9.5 ELECTRONIC ELECTRICITY METERS

Today, electronic electricity meters are used in most developed countries in the world. They have small error limits (below 0.1%), and are generally better than induction meters that were used for most of the 20th century. They also have many other advantages over their analogue counterparts, such as significantly lower consumption, insensitivity to the position of the installation, and less dependency on changes in voltage and frequency.

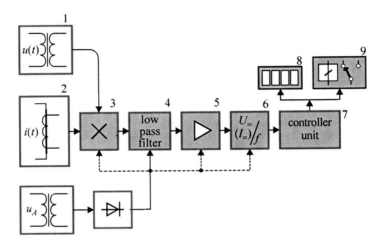

Figure 9.18 Principal circuit of electronic electricity meter

Another advantage of electronic electricity meters is that they have the ability to store data, and can support various automatic meter reading and smart grid technologies, which is especially important when the task is to read many meters in one large building.

The main part of the electronic electricity meter (Figure 9.18) is the measuring transducer that converts the measured power to standardized voltage or current. The converter consists of transformers (1,2) that convert the currents and voltage to the values suitable for multiplier, analogue multiplier (3), and low pass filter (4) that separates the DC output voltage component proportional to the active power of load. After amplification (5), the output voltage is converted into frequency - pulses (6). This procedure is managed by a special controller unit (7). The number of pulses over time is proportional to the energy **$W=Pt$**. The impulses are brought to the counter (8) and the output contacts (9).

Selected bibliography:

1. V. Bego, *"Measurements in Electrical Engineering"*, (in Croatian) Tehnička knjiga, Zagreb, 1990
2. Fluke Electronics, *"Calibration: Philosophy in Practice"*, Second Edition, 1994.
3. M. E. Cage, D. Yu, B. M. Jeckelmann, R. L. Steiner, R. V. Duncan: *"Investigating the Use of Multimeters to Measure Quantized Hall Resistance Standards"*, IEEE Trans. Instrum. Meas. IM-40 2, pp. 262-266., 1991
4. *"HP and Fluke Team to Provide Quick, Affordable Calibration for Popular Digital Multimeters"*, Hewlett Packard Test & Measurement Newsroom, October 14, 1996.
5. S. D. Stever: *"An 8 ½ Digit Multimeter Capable of 100,000 Readings per Second and Two-Source Calibration"*, Hewlett-Packard Journal, pp. 6-21, April 1989.
6. W. C. Goeke, R. L. Swerlein, S. B. Venzke, S. D. Stever: *"Calibration of an 8 ½ Digit Multimeter from Only Two External Standards"*, Hewlett-Packard Journal, pp. 22-49, April 1989.
7. Hewlett-Packard 3458A Multimeter, Operating, Programming and Configuration Manual.
8. A. Sosso, R. Cerri: *"Calibration of Multimeters as Voltage Ratio Standards"*, Conference Digest of CPEM 2000, Sydney, Australia, pp. 375-376.
9. Leniček, I. Ilić, D. Malarić, R.: *"Determination of High-Resolution Digital Voltmeter Input Parameters"*, IEEE Transactions on Instrumentation and Measurement, August 2008, 1685-1688
10. *Make Better AC Measurements with Your Digital Multimeter, Measurement Tips*, Volume 5, Number 2, http://cp.literature.agilent.com/litweb/pdf/5990-3219EN.pdf, Retrieved October 20th, 2010

10. MEASUREMENT OF ELECTRICAL QUANTITIES

There are many quantities by which the behavior of electricity is characterized. Direct-current (DC) measurements include measurements of resistance, voltage, and current in wide ranges and applications, while alternate current (AC) measurements also include measurements of capacitances, inductances, frequencies, and many other quantities. Measurements of electrical quantities extend over a wide dynamic range, with frequencies ranging from 0 to 10^{12} Hz.

10.1 VOLTAGE AND CURRENT MEASUREMENTS

Measurement of voltage and current is performed with instruments of different characteristics; some of these instruments have been described in previous chapters. In every measurement, one should take into account the measuring range, instrument consumption, frequency range, accuracy, and load. Voltmeters are connected to the load in parallel (Figure 10.1.a), while ammeters are connected with the load in series (Figure 10.1.b).

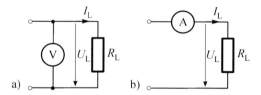

Figure 10.1 Connection of ammeters and voltmeters

The voltmeter consumption must be as low as possible, and its internal impedance must be as large as possible, while the reverse is true for ammeters - their internal impedance should be as small as possible.

10.1.1 Measurement of Small Currents and Voltages

For measuring very small currents and voltages, and also large currents and high voltages, special instruments and methods are required. Currently, sensitive DC voltmeters called **electrometers** are used, enabling the measurement of currents of less than 10^{-17} A. Their input impedance is about 10^{16} Ω. With this instrument, voltages in the order of 10^{-3} V, resistances in the order of 10^{15} Ω, and charge in the order of 10^{-15} C can also be measured. When measuring small currents, the electrometer is measuring the voltage drop on the high-Ohm resistor with a known value (Figure 10.2).

INSTRUMENTATION AND MEASUREMENT IN ELECTRICAL ENGINEERING

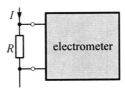

Figure 10.2 Measuring small currents with electrometer

Sensitive **electronic null-instruments** with a sensitive AC amplifier are used for measuring small alternating currents and voltages in the compensation circuits and bridges.

10.1.2 Measurement of Large Currents

Large currents can be measured with the **Hall probe**. The Hall probe (Figure 10.3) consists of a thin plate, through which the control current I_H longitudinally flows.

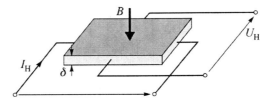

Figure 10.3 Hall probe

If the magnetic field flows perpendicularly to the plate, there is a voltage produced by the magnetic field force on charges in the plate. This voltage, called the Hall voltage, is:

$$U_H = R_H \cdot \frac{BI_H}{\delta} \qquad [10.1]$$

where B is the induction of the magnetic field, and δ the thickness of the plate. The principle of measuring large currents by using Hall probes is shown in figure 10.4.

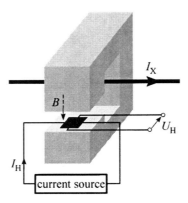

Figure 10.4 Measurement of large currents using the Hall probe

The conductor, through which the measured current I_X flows, is placed in the iron core air gap. The core air gap contains the Hall probe through which a constant control current flows. The field strength in the air gap is proportional to the measured current, so the Hall voltage is also proportional to the measured current:

$$U_H = k \cdot I_X \qquad [10.2]$$

Large currents can be also measured by measuring the voltage drop on shunt and current measuring transformers (Chapter 6).

The field measurements on the transmission network for the distribution of energy are performed by the current clamp (Figure 10.5), which allows current measurement without breaking the circuit.

Figure 10.5 Current clamps

Current clamps are actually a type of a current transformer, with an adjustable core which can enclose a conductor through which the current flows. The conductor is actually one bend of the primary winding, while the secondary windings are wound directly to the core. The secondary circuit contains an ammeter, with which currents of up to several thousand amperes can be measured.

10.1.3 Measurement of High Voltages

For measuring high voltages, there are several methods, some of which have already been described (transformers, capacitive dividers, etc.) In addition to the capacitive divider, resistive (Figure 10.6) and inductive dividers (Figure 10.7) can be used as well.

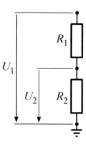

Figure 10.6 Resistance divider for measuring high voltages

A **resistance divider** is a serial connection of two resistors. The measured high voltage is connected to the divider, and the instrument is connected to the low voltage on the resistor R_2. When measuring AC voltage, self capacitance and inductance of the resistors, as well as capacitance towards the earth, must be kept as small as possible.

Small self-inductance and capacitance is achieved by the appropriate winding, while the capacitance towards the earth is reduced by protection rings.

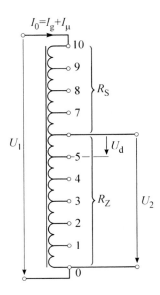

Figure 10.7 Inductive divider

Inductive dividers are voltage transformers with branches that allow the exact voltage division. Figure 10.7 shows the divider in which the connected voltage is divided into ten equal parts. Ten conductors of equal cross-section are first interlocked with each other, and then wrapped around the ring core of high magnetic conductivity. Then they are connected in series, and the connections are provided at joints. Thus, each conductor encloses the same magnetic flux, so the induced voltage of each branch is virtually identical. Because of their good properties, inductive dividers are widely used in compensation circuits together with the Kelvin-Varley divider (Chapter 5.3).

Spherical gaps have been used for a long time to measure peak voltage (Figure 10.8). The lower figure shows a circuit used to test objects at high voltages.

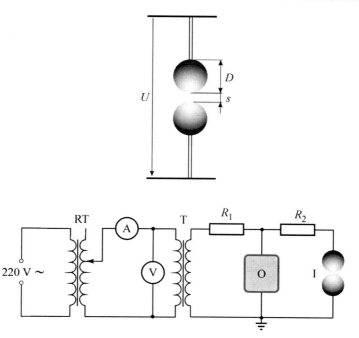

Figure 10.8 Spherical gaps

The measurement is based on the fact that the arcing occurs in the air gap between the spheres, if the peak value of the applied voltage is greater than or equal to some value U_p. During the slow increase of voltage, the arcing will occur exactly when the peak value of voltage will reach the voltage U_p. The arcing voltage depends on the distance of the spheres s, their diameter D, the barometric pressure, and air temperature, and is determined from tables obtained theoretically and experimentally. For small ratios of s/D, there is almost a linear dependence between the arcing voltage U_p and the spheres distance s, while with higher ratios of s/D, the arcing voltage U_p is growing more slowly with the distance s, so the higher the voltages the larger spheres must be used to keep the linear dependence between U_p and s (Figure 10.9).

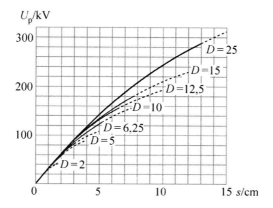

Figure 10.9 Dependence of arcing voltage and the spheres distance

Measurement of high voltages can also be performed by connecting sensitive ammeters in series with the high-value resistor. Voltage is then obtained by multiplying the measured current and the known value of the resistor. It should be understood that the high-ohm resistor must be designed to sustain high voltages.

Peak values of high voltages can be measured by the **Chubb** method (Figure 10.10).

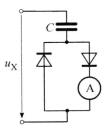

Figure 10.10 **Chubb** method for measurement of high voltages

By applying this method, the peak values of voltage u_X are determined by measuring the average rectified current of the capacitor. The upper electrode of the capacitor **C** is connected to high voltage, and the bottom electrode to the ground through two diodes (semiconductor rectifiers). Diodes are connected in parallel, in the opposite direction of the current flow.

In series with one diode, an instrument that measures the average current value of a half-period is connected, which is proportional to the peak value of the measured voltage (Figure 10.11).

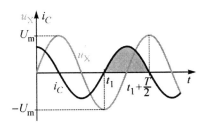

Figure 10.11 Chubb measurement of high voltages, graphic illustration of voltages and currents

The peak value can be calculated as:

$$U_m = \frac{I_{avg}}{2fC} \qquad [10.3]$$

where f is the frequency of the measured voltage.

10.2 POWER MEASUREMENT

The existing power measuring methods differ depending on the parameters in an electrical circuit, particularly the frequency, but also the voltage and power. While the DC power is obtained by multiplying the current and voltage, and is expressed in watts, in AC circuits, this product is the apparent power S, which is expressed in volt-amperes. The apparent power consists of **active** power

$P = UI\cos\varphi$, expressed in watts, and **reactive** power $P = UI\sin\varphi$, expressed in units **var** (volt-ampere reactive). The active and reactive power depends on the phase angle between current and voltage, and instruments that measure them must be phase sensitive. At high frequencies (above 1 GHz), the electric field is proportional to voltage and the magnetic field is proportional to current. While the electric and magnetic fields at high frequencies change depending on the position in the conductor, the power remains the only constant value which is measured by instruments, since the measurement of voltages and currents at these frequencies does not make sense. In the case that the measured quantity is the ratio of two powers, usually the measured power and a reference power, this ratio is usually expressed in decibels, as already discussed in the section on amplifiers (Chapter 7). The reference power is usually a power of 1 mW, and the following formula is used to express power in relative terms:

$$A = 10\log\frac{P}{1\,\mathrm{mW}}\,[\mathrm{dBm}].\qquad[10.4]$$

Then, the power of an oscillator can be converted from dBm to absolute power, so the output power of an amplifier with +13 dBm is 20 mW. The use of decibels has some advantages, as with decibels large ratios can be expressed with several numbers. For example, it is easier to express the power from -153 dBm to +63 dBm than the power from $0{,}5\times10^{-15}$ W to 2×10^{6} W. The other advantage is when making calculations; for example, the total amplification of cascade amplifiers in decibels is summed, not multiplied.

10.2.1 Measurement of DC Power

The DC power can be measured using ammeters and voltmeters. The product of current and voltage gives the power:

$$P = U \cdot I.\qquad[10.5]$$

Instruments can be connected in two different ways. The first option is to connect the ammeter in front of the voltmeter (Figure 10.12), and the second option is to connect the ammeter behind the voltmeter (Figure 10.13).

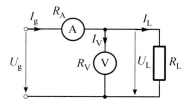

Figure 10.12 Measurement of DC power with ammeter in front of the voltmeter

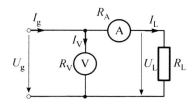

Figure 10.13 Measurement of DC power with ammeter behind the voltmeter

If the power consumed by load is much greater than the consumption of connected instruments, then both ways yield the same results; however, if small powers are measured, the consumption of ammeter and voltmeter must be taken into account. If P_l is the power of load, P_s power of the source, I_g measured current, I_V current of the voltmeter, and R_A and R_V internal impedances of the ammeter and the voltmeter, then the power of load and power of source according to Figure 10.12 is:

$$P_l = U_t I_g - \frac{U_t^2}{R_V}, \qquad [10.6]$$

$$P_s = U_t I_g + I_g^2 R_A. \qquad [10.7]$$

The power of load and power of source according to Figure 10.13 is:

$$P_l = U_g I_t + I_t^2 R_A, \qquad [10.8]$$

$$P_s = U_g I_t + \frac{U_g^2}{R_V}. \qquad [10.9]$$

Which circuit should be chosen depends on the combination of instrument consumption, and the best scheme is where no correction of instrument consumption is necessary. So, when the load resistance is much higher than the ammeter impedance, a scheme according to Figure 10.13 will be better; in the case where the load resistance is negligible in comparison with the voltmeter impedance, a scheme according to Figure 10.12 will be the best option. In the case that a correction cannot be avoided, the selected scheme should be the one in which the correction was needed due to voltmeter consumption, since the voltmeter impedance is known, as opposed to the ammeter impedance which is small in value and usually has temperature dependence. Measurement accuracy depends on the accuracy of the ammeter and voltmeter measurements.

10.2.2 Measurement of Power Using Watt-meters

The power can also be measured using watt-meters, in which case there are also two ways of connecting the instruments (Figures 10.14 and 10.15).

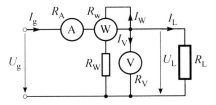

Figure 10.14 Measurement of DC power with ammeter in front of the voltmeter and watt-meter

The determination of source and load power is similar as when using ammeters and voltmeters, but in this case the consumption of current and voltage branches of the watt-meter must be taken into account. The powers of source and load according to Figure 10.14 are:

$$P_1 = P_W - \left(\frac{U_t^2}{R_V} + \frac{U_t^2}{R_W}\right), \qquad [10.10]$$

$$P_s = P_W + I_g^2(R_A + R_W). \qquad [10.11]$$

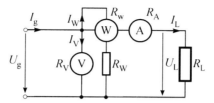

Figure 10.15 Measurement of DC power with ammeter behind the voltmeter and watt-meter

The powers of source and load according to Figure 10.15 are:

$$P_1 = P_W + I_t^2(R_A + R_W), \qquad [10.12]$$

$$P_s = P_W - \left(\frac{U_g^2}{R_V} + \frac{U_g^2}{R_W}\right). \qquad [10.13]$$

When using wattmeters, special care must be taken not to overload its current and voltage branches, as the wattmeter shows the same value for current $I/10$ and voltage $10U$, as well as for current $10I$, and voltage $U/10$. Also, when measuring power with low power factor, despite the large current, the wattmeter will have low deflection. Such situations can result in watt-meter damage, so it is recommended to control the voltage with a voltmeter and the current with an ammeter.

In low-frequency applications, the power is determined by the **electrodynamic** or **electronic** wattmeter. One must be careful when choosing the watt-meter. The power factor is the ratio between **active** and **apparent** power:

$$\cos \varphi = \frac{P}{S}. \qquad [10.14]$$

To measure the power of load with low power factor, wattmeters that have full deflection with nominal currents and voltages and power factor $\cos\varphi=1$ are not suitable, but wattmeters that have full deflection with a lower power factor (e.g. $\cos\varphi=0,2$ ili $\cos\varphi=0,1$) should be used instead.

10.2.3 Connecting the Wattmeter
Wattmeters can be connected directly, semi-directly, and indirectly. Direct connections are used for the measurement of small powers. The voltage measurement range of up to 750 V is extended by serially added resistors. For higher voltage and current ranges, transformers must be used.

INSTRUMENTATION AND MEASUREMENT IN ELECTRICAL ENGINEERING

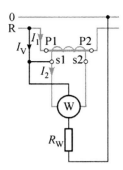

Figure 10.16 Semi-direct connection of the wattmeter

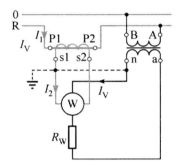

Figure 10.17 Indirect connection of the wattmeter

Semi-direct connection (Figure 10.16.) is used in low voltage applications. Indirect connection is used in high voltage systems (Figure 10.17). On the secondary windings of transformers, there must be an equipotential connection connected to the grounding wire to avoid a large potential difference between the current and voltage coil.

Power measurements can be done without the wattmeters if only voltmeters or ammeters are available. Although in practice such cases are rare, the three voltmeter and three ammeter methods are interesting because of the mathematics by which the active power and power factor of load is calculated.

10.2.4 Three Voltmeter Method

According to Figure 10.18, in a series with load Z_L, which is generally a combination of resistance and inductance, a "pure" resistor R must be connected, which is preferably of the same value as the load resistance.

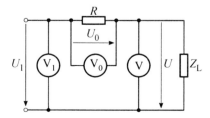

Figure 10.18 Three voltmeter method

MEASUREMENT OF ELECTRICAL QUANTITIES

With an available voltmeter, voltage on Z_L, the **R** and the voltage on the serial connection of two resistors are measured. From the vector diagram (Figure 10.19), by using the cosine theorem, active power can be obtained:

$$P = \frac{U_1^2 - U_0^2 - U^2}{2R}.$$ [10.15]

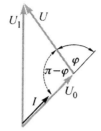

Figure 10.19 Vector diagram of the three voltmeter method

The power factor can be also calculated:

$$\cos\varphi = \frac{U_1^2 - U_0^2 - U^2}{2U_0 U}.$$ [10.16]

The method depends on the voltmeter accuracy and its consumption, so it is suitable for measuring powers of small resistances. In addition, the method is applicable in high frequency ranges, by using electronic voltmeters with a large input resistance. This method is also used for highly accurate power measurements, if the applied voltmeters have narrow limits of error.

10.2.5 Three Ammeters Method

This method is similar to the three voltmeter method, but in this method, the known resistor **R** is connected parallel to Z_L (Figure 10.20).

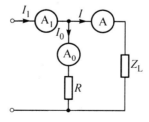

Figure 10.20 Three ammeter method

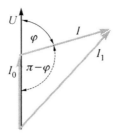

Figure 10.21 Vector diagram of the three ammeter method

According to the vector diagram (Figure 10.21), the active power can be determined:

$$P = R \frac{I_1^2 - I_0^2 - I^2}{2}. \qquad [10.17]$$

The power factor of load can also be calculated:

$$\cos \varphi = \frac{I_1^2 - I_0^2 - I^2}{2 I_0 I}. \qquad [10.18]$$

It is necessary that the ammeter internal resistances be as low as possible compared to the load resistance, which makes this method suitable for measuring the powers of large resistances.

10.2.6 Measurement of Active Power in Three-Phase Systems

The three-phase system can be divided into the **three-wire** system without neutral and **four-wire** system with neutral. The three-wire system is usually used in high voltage transmission systems and the four-wire system is used in low voltage networks. Three phases are identified by the letters **R**, **S**, and **T**. The power of the three-phase symmetric system (symmetric meaning that the loads are equal in all phases) with neutral can be measured by measuring the power of one phase (e.g., phase R) multiplied by three:

$$P = 3 \cdot P_R = 3 \cdot U_R \cdot I_R \cdot \cos \varphi_R. \qquad [10.19]$$

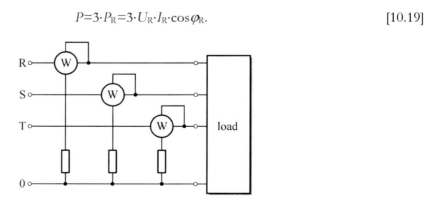

Figure 10.22 Power measurement of three phase assymmetric system

MEASUREMENT OF ELECTRICAL QUANTITIES

The power of the three-phase assymmetric system is calculated as the sum of the powers of each of the three phases (Figure 10.22). Wattmeter current coils are connected to each of the phase conductors, and voltage coils are connected to the phase voltages, i.e., between the phase and neutral conductors.

The total power and phase powers are calculated according to [10.20-10.23]:

$$P_R = U_R \cdot I_R \cdot \cos\varphi_R \quad [10.20]$$

$$P_S = U_S \cdot I_S \cdot \cos\varphi_S \quad [10.21]$$

$$P_T = U_T \cdot I_T \cdot \cos\varphi_T \quad [10.22]$$

$$P_{tot} = P_R + P_S + P_T \quad [10.23]$$

where U_R, U_S i U_T are phase voltages, and I_R, I_S i I_T are currents in phase conductors.

Wattmeters can be connected directly and semi-directly, in analogy with one-phase systems.

The power of the three-wire three-phase system is measured so that the output voltage terminals are connected together and thus establish an artificial null of the system. Measurement of the symmetrical system can be done by measuring the power in only one phase, provided that the voltage branches of the other two phases are replaced with two resistors of equal value to the resistance R_W (Figure 10.23).

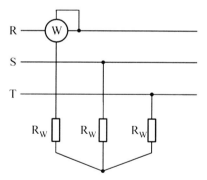

Figure 10.23 Measurement of three phase symmetric system using only one wattmeter

In the assymmetric system, it is necessary to measure the powers in all three phases. It is also necessary that the wattmeter voltmeter branch resistance R_W is equal in all phases (Figure 10.24).

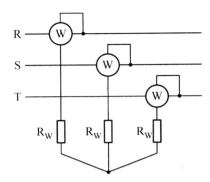

Figure 10.24 Measurement of three phase assymmetric system using three wattmeters

10.2.7 Aron Connection

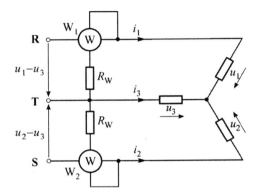

Figure 10.25 Three phase power measurement using Aron connection

The power of the three-phase system without neutral, both symmetrical and assymmetric can be measured by only two wattmeters using the **Aron connection**. Perhaps there was a shortage of wattmeters during Aron's time - this method is interesting because it displays the engineer's ingenuity. Current branches of wattmeters **W₁** and **W₂** are connected in two phases, and voltage branches are connected to the third phase, where there is no current branch (Figure 10.25).

Figure 10.26 shows the vector diagram of **three-phase symmetric system**, in which the phase currents lag behind the phase voltage by angle φ.

MEASUREMENT OF ELECTRICAL QUANTITIES

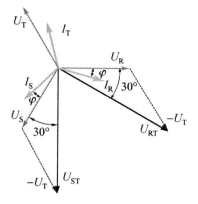

Figure 10.26 Vector diagram of the Aron connection

The voltage branch of wattmeter W_1 is affected by line voltage U_R-U_T and current I_R between which there is a phase shift of 30°-φ, so the wattmeter W_1 will show:

$$P_1 = I_R U_{RT} \cos(30° - \varphi) \qquad [10.24]$$

According to the same vector diagram, the wattmeter W_2 will show:

$$P_2 = I_S U_{ST} \cos(30° + \varphi) \qquad [10.25]$$

The sum of the first and second wattmeters will give the mean power of all three phases:

$$P = P_1 + P_2 \qquad [10.26]$$

Wattmeter readings of each wattmeter do not make sense, because one wattmeter may show "negative" power with reactive loads when the phase angle is greater than 60°. If this occurs, the voltage connections terminal of that wattmeter must be replaced, and the reading of that wattmeter is taken with a negative sign. Accuracy is higher as the power factor of load is higher, and this method is not suitable for power factors of less than 0.3. The power factor can also be calculated knowing two voltmeter readings. If $\xi = \dfrac{P_2}{P_1}$ holds, the power factor is obtained as:

$$\cos\varphi = \dfrac{1}{\sqrt{1 + 3\left(\dfrac{1-\xi}{1+\xi}\right)^2}} . \qquad [10.27]$$

10.3 RESISTANCE MEASUREMENT

Measurement of resistance requires methods for measuring very small to very large resistances. A large number of non-electric quantities are also reduced to the measurement of resistance (e.g., temperature). Therefore, for the measurement of resistances, there are many more measurement methods available than for any other quantity.

Before the measurement of resistance can begin, some facts must be known in advance. The conductors used to connect the measured resistance to the instrument also have some resistance which adds to the measured resistance. Also, **contact resistances** can cause errors in the measurement of small resistances, as contact resistances can have values similar to the measured resistance.

In addition, if the place where two different metals or alloys touch each other is heated, then **thermoelectric voltage** is produced at the free end of these metals. This voltage depends on the temperature difference between the hot and cold junction and causes the error. Therefore, thermoelectric voltages must be kept as low as possible on the connection terminals of the resistors. The temperature coefficient of resistors differs from material to material, and for accurate measurements, the temperature needs to be measured and corrections made. Some resistors also exhibit voltage resistance dependence.

10.3.1 Voltmeter-Ammeter Method

One of the simplest methods of indirect measurement of resistance is the voltmeter-ammeter method.

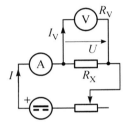

Figure 10.27 Voltmeter-ammeter "voltage method" for measuring resistances

According to this method, the resistance is obtained from the quotient of voltage drop on resistance R_X and the current flowing through the resistor. At low resistance values, the voltage method (Figure 10.27) is used because a voltmeter with high internal resistance consumes negligible power. At high value resistances, the current method (Figure 10.28) is used, as in this method, small resistance of ammeters can be neglected.

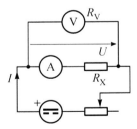

Figure 10.28 Voltmeter-ammeter "current method" for measuring resistances

The digital ohm-meter is based on this principle, as it passes a "known" current through the resistor and then measures the voltage drop on it.

MEASUREMENT OF ELECTRICAL QUANTITIES

10.3.2 Compensation and Digital Voltmeter Methods

According to this method, two resistors are connected in series, where the unknown resistance is compared with a known resistance standard. A known current is passed through these two resistors and the voltage drop is measured on each resistor (Figure 10.29).

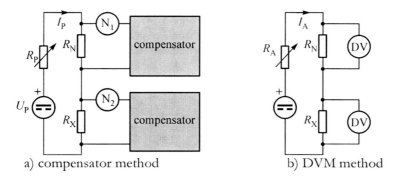

a) compensator method   b) DVM method

Figure 10.29 Compensating methods for the measurement of resistance standards

The voltage drop on each resistor is measured by placing the resistor into opposition to the known voltage which is changed manually or automatically until the null-detector remains without deflection (Figure 10.29a). From this state, an unknown voltage drop is equal to the known voltage, thus the unknown voltage can be determined. The resistance ratio is proportional to the voltage ratio and the unknown resistance can be easily calculated by multiplying the "known" resistance value of the resistance standard with the voltage ratio:

$$R_1 = R_2 \frac{U_1}{U_2} \qquad [10.28]$$

This method requires a stable voltage source because its stability directly affects measurement accuracy. In equilibrium, no current flows through the circuit. Recently, high accuracy digital voltmeters with high input impedance have replaced the compensator in this method (Figure 10.30).

The Primary Electromagnetic Laboratory (PEL) at FER-Zagreb is calibrating the resistance standards using the DVM comparison method to calibrate the resistance standards from 1 mΩ up to 100 MΩ (Figure A1.1). These methods were validated and accredited by the German accreditation service Deutsche Kalibrier Dienst (DKD) in 2005. The traceability chain of the PEL resistance standards are presented in Fig. 3.11. Apart from the accreditation of resistance standards that span from 1 mΩ to 100 MΩ, the laboratory is accredited for the voltages from 1 V to 10 V, and capacitance standards of 100 pF. The total relative uncertainty (k=2) for the whole range of resistance standards does not exceed $12 \cdot 10^{-6}$, while for the standards in the range from 100 mΩ to 1 kΩ, the relative uncertainty is $1 \cdot 10^{-6}$. Three standards (1 Ω L&N 4210, 10kΩ L&N 4040-B, and 10 MΩ Fluke 742A) are calibrated against the Quantum Hall resistance standard at PTB every three years. The rest of the resistance standards are traced to these three standards with the DVM comparison method. Some of the resistance standards that have high temperature coefficients are placed in a thermostated oil bath that is kept at temperatures of 23 °C ± 0,03 °C (Figures 3.9 and 3.10), while the rest are kept in a regulated laboratory temperature of 23 °C ± 0,5 °C. Most of the standards have well-preserved historical calibration data of more than 30 years, enabling the good forecasting of resistance values between international calibrations.

The resistance standards are compared using two digital voltmeters. The whole measurement process is automatized using personal computer and LabVIEW software (Figure A1.2). The front panel contains all the necessary data for the measurement report, including measurement setup, results, time and date, and standards temperature, while the results are presented graphically during the measurement. The data rejection methods can be used to reject the data that are off limits.

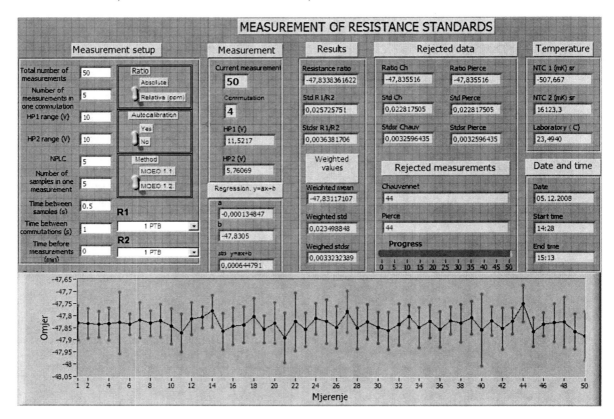

Figure 10.30 LabVIEW front panel for the measurement of resistance standards

The standard deviation of such comparison is usually less than $1 \cdot 10^{-8}$. Each ratio consists of four measurements, as DVMs are interconnected to each resistance and the current is reversed with the help of a specially developed relay box. In the end, the weighted mean and standard deviation can be calculated which usually gives better results than a normal average ratio. After a ratio is determined, then the standard resistance value of unknown standard is easily calculated using the known value of the resistance standard used for comparison.

10.3.3 Measuring Resistance Using the Ohm-Meter

Ohm-meters are measurement instruments which directly indicate the value of the measured resistance. As a measuring instrument, the most used is an instrument with moving coil and a permanent magnet. According to Ohm's law, the current is inversely proportional to the resistance of the

MEASUREMENT OF ELECTRICAL QUANTITIES

resistor in a circuit with constant voltage, so the instrument can be scaled in Ohms, if constant voltage or current is maintained. The simplified method is shown in Figure 10.31.

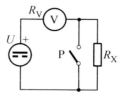

Figure 10.31 Ohm-meter method

The voltmeter with internal resistance R_V is connected in series with the measured resistance R_X, which can be short-circuited with button **P**. The instrument (ammeter) deflects α_1 with the open button, and α_2 with the closed button, so the currents are:

$$I_1 = \frac{U}{R_V} = k\alpha_1 \qquad [10.29]$$

$$I_2 = \frac{U}{R_V + R_X} = k\alpha_2 \qquad [10.30]$$

From [10.29] and [10.30], unknown resistance R_X can be calculated:

$$R_X = R_V \left(\frac{\alpha_1}{\alpha_2} - 1 \right). \qquad [10.31]$$

The deflection of the instrument depends only on the deflection ratio and internal resistance of the instrument. Deflection α_1 is chosen to be as large as possible, so it is usually a full deflection of the instrument. In this way, the measurement of the resistance will be reduced to performing only one measurement, i.e., α_2.

10.3.4 Digital Ohm-Meter

There are two basic types of digital ohm-meters:
- successive approximation digital ohm-meter (chapter 9.1.3)
- current source digital ohm-meter

Ohm-meters with successive approximation consist of a Wheatstone bridge, where the unknown resistor is placed in one branch; the second branch consists of a resistor decade used for automatic balancing by a control device. Resistances in the other two branches are used for range selection of the measuring instrument.

Digital ohm-meters can also be derived from **digital voltmeters** with a current source; the measurement of resistance is reduced to measuring the voltage drop on the unknown resistor using the known current I:

$$R_X = \frac{U_X}{I} \qquad [10.32]$$

10.3.5 Measurement of Insulation Resistance

It is important to regularly perform insulation resistance measurements to ensure the maintenance of electrical equipment. It is estimated that nearly 80% of all maintenance activities in the industry is related to checking the insulation of machines. Measurement of insulation resistance, which amounts to several hundred megaohms, requires high voltages of several hundred volts, and is performed by using Ohm's Law. Since such measurements are often carried out in the field, the instrument for measuring insulation resistance has a small DC voltage generator, which is converted into a few hundred volts by using the appropriate converter. Insulation meters are usually built around an instrument with a moving coil and permanent magnet which can have an LCD display. Insulation resistance ages and is also dependent on temperature, duration of measurement, humidity, and many other influential quantities; therefore, a high reproducibility of results cannot be achieved. The result of measurement is then stated in such a way that the insulation resistance is greater than or equal to some value, e.g., $R_X \geq 100\,\text{M}\Omega$.

10.3.6 Measurement of High-Ohm Resistance

Measurement of high resistance is performed by:
a) Megaohm-meters, devices that have high input impedance (e.g. electrometers)
b) The discharge method – capacitor **C** is charged to the known voltage and then the time **t** is measured, during which capacitor is discharged through the unknown resistor to half of its initial value (Figure 10.32).

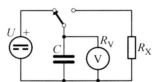

Figure 10.32 Measurement of high resistances by the capacitor discharge method

The voltage on the capacitor is $U_c = U \cdot e^{-t/RC}$, so the resistance can be calculated:

$$R = \frac{\Delta t}{C \cdot \ln \dfrac{U}{U_c}} \qquad [10.33]$$

c) A specially adapted Wheatstone bridge, where various insulation resistances are taken into account, which may interfere with the measurement.
d) The voltmeter-ammeter method, which is used to measure the surface resistances and conductivity of insulating samples (Figures 10.33 and 10.34). The sample is placed between electrodes **1** and **3**, and the top electrode is connected to the measuring instrument and the voltage source to the lower electrode. Around the upper electrode, is a grounded protective electrode **2** to eliminate surface currents.

MEASUREMENT OF ELECTRICAL QUANTITIES

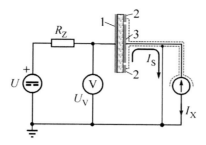

Figure 10.33 Circuit for measuring the insulation resistance of sample

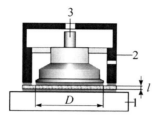

Figure 10.34 Insulation material sample setup for measurement

The measurement of high resistances must be carefully performed (Figure 10.35). The insulation resistance of the stand at which the resistor is placed can be of the same magnitude as the measured resistor, so the measured resistance is in fact a parallel combination of the test resistor and insulation resistance. This problem can be solved by shielding or protective electrodes. Thus, the surface currents flowing through the insulation resistance are not flowing through the measuring instrument. Figure 10.35 shows a test stand with the high-Ohm resistor. The microammeter in the voltmeter-ammeter method will measure both the current I_X through the test resistor R_X and current I_S through the insulation resistance, causing a large measurement error.

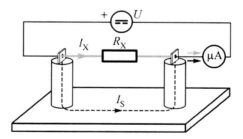

Figure 10.35 Measurement of high-Ohm resistors with the voltmeter-ammeter method

If there is a protective conductive ring placed around one column and if an ammeter is connected to that ring, the ammeter will measure only current I_X through the test resistor (Figure 10.36).

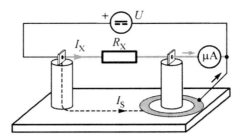

Figure 10.36 Protective ring placed around one of the columns holding the test resistor

10.3.7 Measurement of Earth Resistance

All parts of electrical equipment that do not belong to the circuit, and are exposed to contact, must be grounded. Earth resistance must be low enough that in the case of equipment failure, which can be life-threatening, large currents can flow which will quickly melt the protective fuse and thereby interrupt the circuit.

The grounding electrode may be of different construction, and for large facilities rings, grilles, and additional electrodes can be used. The potential away from the electrode falls as seen in Figure 10.37.

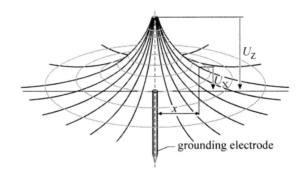

Figure 10.37 Fall of potential from the grounding electrode

The usual method for determining earth resistance is the "fall of potential method," as seen in Figure 10.38. The resistance of an earthing electrode is determined by passing the AC current through the electrode and then measuring the voltage drop on it. Therefore, it is necessary to place an additional probe P_2 at sufficient distance from the earthing electrode Z and connect the voltage source U between the electrode and the probe. Current density will be greatest along the earthing electrode Z and probe P_2, so for this reason it is necessary to place another probe P_1 between the earthing electrode Z and the probe P_2 (Figure 10.38).

MEASUREMENT OF ELECTRICAL QUANTITIES

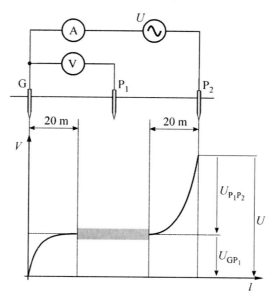

Figure 10.38 Earth resistance measurement method

Current density at the place of probe P_1 must be negligible, so the voltmeter which is connected between the P_1 probe and earthing electrode Z, will measure the voltage drop only on the earthing electrode Z. The P_1 probe must be at least 20 meters away from the earthing electrode Z, but also 20 meters away from the probe P_1. Then the earth resistance is calculated as:

$$R_G = \frac{U_{ZP1}}{I}. \qquad [10.34]$$

10.3.8 Measurement of Soil Resistivity

The measurement of soil resistivity is necessary for designing the dimensions of the earthing electrode. The simplest measurement of soil resistivity is performed with four electrodes stuck into the earth at large enough distances. The voltage source is connected to external electrodes, and the voltage drop U_{BC} is measured on the internal electrodes (Figure 10.39).

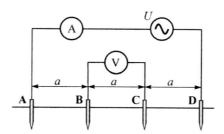

Figure 10.39 Measurement of soil resistivity

Soil resistivity is then calculated as:

157

$$\rho = 2a\pi \frac{U_{BC}}{I}. \qquad [10.35]$$

10.4 MEASUREMENT OF IMPEDANCE

Impedance **Z** is usually a serial and/or parallel combination of real or active (resistance R) components and imaginary or reactive components, usually expressed in ohms (Ω). The reactive component is called **reactance**, denoted with **X**, and it can be **inductive** $(X_L = \omega L)$ or **capacitive** $(X_C = 1/\omega C)$. For the parallel combination of resistance and reactance, **admittance Y** is often used for convenience of writing, and it is the reciprocal of impedance, expressed in **siemens** (**S**). Admittance has two components, real or **conductance G**, and reactive or **susceptance B**. For coils, the important quantity is the quality factor **Q=X_L/R**, and for capacitors, the important quantity is the loss tangent (**tan δ**). It is desirable that the coils have a higher Q-factor and the capacitors have a small loss tangent. When measuring real coils and capacitors, the equivalent resistance should be taken into account (R_S - equivalent serial resistance and R_p - parallel equivalent resistance). Depending on the application, different methods are used to measure the impedances, such as the voltmeter-ammeter method, comparisons, bridge, etc. It is also important that both real and imaginary components are measured separately. In the upcoming chapters, different methods of measuring inductance and capacitance will be described, while in Chapter 10.8, a digital self-adjustment instrument, which can measure two or more impedance quantities (Z, φ, Y, R, X, B, G, Q, tg δ) within a very short time, will be described.

10.5 MEASUREMENT OF INDUCTANCE

Inductance coils without iron cores can be characterised by measuring the voltage and current (Figure 10.40).

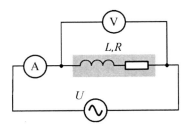

Figure 10.40 Measurement of coil inductance with the voltmeter-ammeter method

If the AC voltage is applied across the coil, the current can be calculated as:

$$I = \frac{U}{\sqrt{R^2 + (\omega L)^2}}. \qquad [10.36]$$

The term in the denominator is the coil impedance:

$$Z = \sqrt{R^2 + (\omega L)^2}. \qquad [10.37]$$

If resistance R of the coil is determined by one of the methods for measuring resistance, then the inductance of the coil can be determined if the frequency of AC voltage is also known or measured:

$$L = \frac{\sqrt{Z^2 + R^2}}{\omega}.\qquad [10.38]$$

This procedure can be used only if the losses in the coil in the DC and AC currents are equal. As the losses occur in the iron core with alternating current, this method cannot be used for coils with an iron core. Then the method according to Figure 10.41 is used.

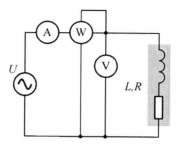

Figure 10.41 Wattmeter-voltmeter-ammeter method for measuring inductance

Current and voltage of coil are measured with ammeters and voltmeters, and the total losses *P* with the wattmeter. Then the inductance is calculated:

$$L = \frac{1}{\omega I^2}\sqrt{U^2 I^2 - P^2}.\qquad [10.39]$$

10.5.1 Bridge with Variable Inductance

Inductance and associated resistance in the bridge methods are compared with the knself-inductance or capacitance, and active resistances. In the first branch of the bridge with variable inductance (Figure 10.42), there is the coil with unknown self-inductance L_X and active resistance R_X.

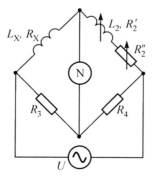

Figure 10.42 Bridge with variable inductance

The other branch has variable inductance coil L_2, active resistance R_2', and another variable resistor R_2''. In the third and fourth branches, there are the resistors R_3 and R_4. So, in all branches of the bridge, there are the following impedances:

$$Z_1 = (R_x + j\omega L_x) \quad [10.40]$$
$$Z_2 = (R_2' + R_2'' + j\omega L_2) \quad [10.41]$$
$$Z_3 = R_3 \quad [10.42]$$
$$Z_4 = R_4 \quad [10.43]$$

For a balanced bridge the following is true: $Z_1Z_4=Z_2Z_3$, and the unknself-inductance and resistance after multiplication and separation of real and imaginary components can be calculated:

$$R_X = \frac{R_2 R_3}{R_4} \quad [10.44]$$

$$L_X = \frac{L_2 R_3}{R_4}. \quad [10.45]$$

The bridge is balanced with the variable resistor R_2'' and inductance L_2, and their balancing is mutually independent, so the balance of the bridge, at least theoretically, can be achieved with the minimum number of tuning (two). Variable inductance is an element that is not always available, so capacitor and resistance decades are used for balancing purposes instead.

10.5.2 Maxwell Bridge
Figure 10.43 shows the Maxwell bridge, which has the following impedances:

$$Z_1 = (R_x + j\omega L_x) \quad [10.46]$$
$$Z_2 = R_2 \quad [10.47]$$
$$Z_3 = R_3 \quad [10.48]$$
$$Z_4 = \frac{R_4}{1 + j\omega C_4 R_4}. \quad [10.49]$$

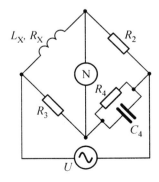

Figure 10.43 Maxwell bridge

The balance of the bridge is achieved with independent tuning of C_4 and R_4; however, the balancing elements can be also R_4 and R_2 or R_3, when capacitor decades are not available. Once the bridge is balanced, it follows:

$$R_x = \frac{R_2 R_3}{R_4} \qquad [10.50]$$

$$L_x = C_4 R_2 R_3 . \qquad [10.51]$$

10.6 CAPACITANCE MEASUREMENT

Just like for inductance, measuring capacitance can be performed by measuring current, voltage, and frequency as shown in Figure 10.44.

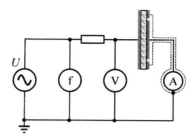

Figure 10.44 Measurement of capacitance with the voltmeter-ammeter method

The capacitance is then:

$$C = \frac{I}{2\pi f U} \qquad [10.52]$$

The above formula applies only to capacitors with negligible losses and sinusoidal signals, because at higher harmonics, the reactance of the capacitor is no longer $X_C = \dfrac{1}{2\pi f C}$.

10.6.1 Wien Bridge

The capacitor to be measured is placed in the first branch and presented with a serial combination of C_x and R_X. (Figure 10.45).

The second branch contains a variable capacitor and variable resistor through which an independent balancing is achieved, while the third and fourth branches contain resistors. The impedances of the branches are:

$$Z_1 = \left(R_x + \frac{1}{j\omega C_x} \right) \qquad [10.53]$$

$$Z_2 = \left(R_2 + \frac{1}{j\omega C_2} \right) \qquad [10.54]$$

INSTRUMENTATION AND MEASUREMENT IN ELECTRICAL ENGINEERING

$$Z_3 = R_3 \qquad [10.55]$$
$$Z_4 = R_4. \qquad [10.56]$$

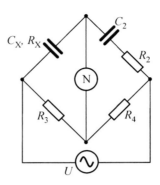

Figure 10.45 Wien bridge

After separating real and imaginary components from [10.53-10.56], the C_x and R_x can be calculated:

$$R_x = \frac{R_2 R_3}{R_4} \qquad [10.57]$$

$$C_x = C_2 \frac{R_4}{R_3}. \qquad [10.58]$$

Good adjustment can be achieved by changing the R_2 and R_3, or R_2 and R_4. The Wien bridge is most suitable for measurement at low voltages, while for higher voltages, the Schering bridge must be used.

10.6.2 Schering Bridge

The Schering bridge is used for measuring the **loss angle** of insulation materials and electrical equipment, especially at high voltages. High-voltage branches are made of measured capacitor C_x (e.g., high voltage cable sample) and standard capacitor C_2 (Figure 10.46).

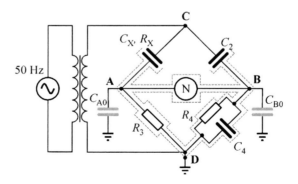

Figure 10.46 Schering bridge

The bridge is balanced by elements in the third and fourth branches of the bridge, and these elements are chosen in such a way that the voltage drops on them are much lower than the high voltages on C_X and C_2.

Impedances of the individual branches in the bridge are:

$$Z_1 = \left(R_x + \frac{1}{j\omega C_x}\right) \tag{10.59}$$

$$Z_2 = \frac{1}{j\omega C_2} \tag{10.60}$$

$$Z_3 = R_3 \tag{10.61}$$

$$Z_4 = \left(\frac{R_4}{1 + j\omega C_4 R_4}\right) \tag{10.62}$$

The bridge is balanced with R_3 and C_4. Resistors R_3 and R_4 are selected so that practically all voltage remains on C_x and C_2. This ensures that balancing elements R_3 and C_4 and the null detector are at negligible voltages towards the ground, which allows safe handling of the bridge during measurement. After balancing and separating real and imaginary parts, C_X and R_X can be calculated:

$$C_X = C_2 \frac{R_4}{R_3} \tag{10.63}$$

$$R_x = \frac{C_4}{C_2} R_3. \tag{10.64}$$

The loss tangent is:

$$\text{tg}\delta = \omega C_X R_X = \omega C_4 R_4. \tag{10.65}$$

Stray capacitances towards earth C_{A0} and C_{B0} are parallel to the third and fourth branches of the bridge, and they directly influence the measurement results. Since they are very unstable due to environmental influences, they must be stabilized and taken into account. This is achieved by a **shielding** enclosure of the lower sections of the bridge (indicated by dotted lines in Figure 10.46) and measuring these capacitances towards the grounded shield. The loss tangent can be then calculated as:

$$\text{tg}\delta = \omega[R_4(C_4 + C_{B0}) - R_3 C_{A0}] \tag{10.66}$$

A commercial Schering bridge can be seen in Figure 10.47.

Figure 10.47 Commercial Schering bridge

Stray capacitance errors are more severe if one side of the AC supply is grounded. The effect of stray capacitances can be reduced if the null detector is kept at ground potential, so there will be no AC voltage between it and the ground, and thus no current through stray capacitances. However, directly connecting the null detector to the ground would create a direct current path for stray currents. The solution is then to use a special voltage divider circuit called Wagner ground or Wagner auxiliary bridge to maintain the null detector at ground potential without the need for a direct connection to the null detector (Figure 10.48), thus minimizing the effects of stray capacitance to ground on the bridge.

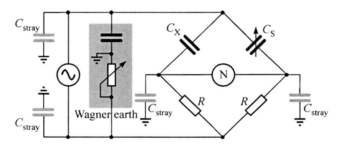

Figure 10.48 Wagner auxiliary bridge

The Wagner auxiliary bridge is composed of a voltage divider designed to have the voltage ratio and phase shift as each side of the bridge. As the midpoint of the Wagner divider is directly grounded, any position in the main bridge circuit having the same voltage and phase as the Wagner divider, and powered by the same AC voltage source, will be at ground potential as well (also called virtual ground). Thus, the Wagner earth divider forces the null detector to be at ground potential, without a direct connection between the detector and the ground.

There is often a two-position switch made in the null detector connection to confirm proper setting (Figure 10.49), so that one end of the null detector may be connected to either the main bridge or the Wagner bridge.

MEASUREMENT OF ELECTRICAL QUANTITIES

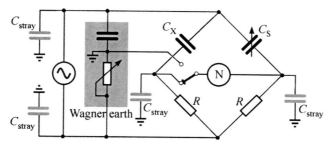

Figure 10.49 Switch up position of null detector

When the null detector shows zero in both switch positions, the bridge is balanced, but at the same time also at zero potential with respect to the ground, thus eliminating any errors due to leakage currents through stray capacitances.

10.6.3 Transformer Bridges

Capacitance can be measured using the transformer bridge, where a special current transformer with branches is placed in the lower section of the bridge. Figure 10.50 shows the simplified **Glynn** transformer bridge, which uses a current transformer with three windings.

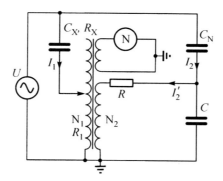

Figure 10.50 Glynn transformer bridge

Through the first winding with many branches that can be connected using the switch, the current of the measured capacitance C_x flows. The second winding has several branches and through it the standard capacitor C_N current flows. A null detector is connected to the third winding. If balancing is done by changing the number of turns of the first winding, until the product of the first and second winding turns and currents is equal ($N_1 I_1 = N_2 I_2$) and in opposite direction, which is achieved by opposite directions of the winding, there will be no magnetization of the transformer core and induced voltage in the third winding, so the null detector will show zero deflection. Then, through the first winding current $I_1 = U C_X \omega$ will flow, and in the second winding current $I_2 = U C_N \omega$ will flow, so the measured capacitance can be calculated as:

$$C_X = C_N \frac{N_2}{N_1}. \qquad [10.67]$$

10.7 MEASURING IMPEDANCE BY SELF-ADJUSTING BRIDGE

The principal circuit of a digital LCR meter is shown in Figure 10.51; with this instrument, resistance, capacitances, inductances, loss angle, factor, and many other quantities can be measured.

A principle of measurement will be explained in measuring the capacitance C_x presented with equivalent capacitance C_p and resistance R_p. A sine voltage is applied to C_p and R_p, causing the current flow through them to the inverting terminal of the operational amplifier (connected as a current to the voltage converter). The total current passing through the capacitor is the sum of capacitive and real current, as shown in Figure 10.52.

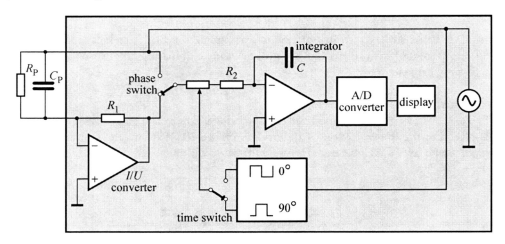

Figure 10.51 Principal circuit of digital LCR meter

The real current is flowing through the equivalent resistance R_p and capacitive current is flowing through the equivalent capacitor C_p. The feedback current to the inverting terminal from the output of the operational amplifier is phase-shifted by 180°.

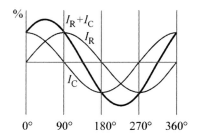

Figure 10.52 Current through the capacitor (the sum of resistive and capacitive currents)

The output voltage of the operational amplifier is the product of two currents and the feedback resistor. The output of the operational amplifier (the current to voltage converter) or the output of the sine source can be passed on to the integrating amplifier by the phase switch. The phase switch is opened by the rectangular time signal, which is either in phase with the sine voltage source or 90° phase-shifted from it. The output of the integrating amplifier is converted into digital form in analogue to digital converters, and displayed on the instrument. In determining the capacitance C_P, the output of current to voltage converter is applied to the integrating amplifier when the rectangular time signal opens the phase switch. In Figure 10.53, this section is unshaded.

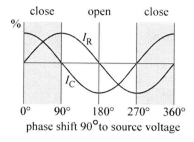

Figure 10.53 Time signal phase shifted by 90° from the voltage source

The mean value of active current I_R, which passes through the equivalent capacitance in this half-wave, is therefore zero, so the integrated value of the current in this period will also be zero. The mean value of capacitive current in the same half-wave is obtained as a DC charge on the integrating amplifier. After the capacitor is charged, the DC voltage on the feedback's capacitor is $U=-U_i\omega C_P T_1$, where U_i is the source voltage and T_1 is the integration time. After T_1, the input to the integrating amplifier is obtained from the voltage source when the rectangular time signal in phase with the voltage source opens a phase switch. This causes the capacitor voltage to discharge to zero volts. The time of discharge is T_2, so for the whole process of charging/discharging of the capacitor, the following applies:

$$-U_i \cdot \omega \cdot C_P \cdot T_1 = U_i \cdot T_2 \qquad [10.68]$$

Then the equivalent capacitance C_P can be calculated:

$$C_P = \frac{T_2}{\omega T_1} \qquad [10.69]$$

To measure the loss angle of the capacitor, the output of the current to voltage converter is applied to the integrating amplifier when the rectangular time signal in phase with the voltage source opens the phase switch. In Figure 10.54, the input of the integrating amplifier is unshaded. The value of capacitive current I_C, which passes through the capacitor in this half-wave is zero, so the integrated value of the current will also be zero.

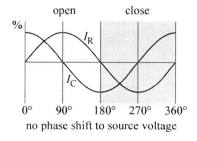

Figure 10.54 Time signal in phase with the voltage source

The mean value of active current I_R through the equivalent resistance of the capacitor R_P in the same half-wave is obtained as a DC charge of the integrating amplifier. When charging is done, the DC voltage on the feedback's capacitor is $U=-U_i T_3/R_P$, where U_i is the source voltage and T_3 the integration time. After the integration period T_3, the capacitive current I_c is applied to the integrating amplifier when the source opens the phase switch. This causes the integrating capacitor to discharge to zero volts. The discharge time is T_4, so for the whole process of charging/discharging the capacitor, the following applies:

$$-U_i \cdot T_3 / R_P = -U_i \cdot \omega \cdot C_P \cdot T_4 \qquad [10.70]$$

Then the loss angle of the capacitor is:

$$\tan \delta = \frac{1}{R_p C_p \omega} = \frac{T_4}{T_3} \qquad [10.71]$$

10.8 TIME, FREQUENCY, AND PERIOD MEASUREMENTS

Time can be measured electronically with the simple method presented in Figure 10.55. The impulse generator of stable and known frequency f_N (START and STOP) is controlled by the measured quantity. The time t is determined from the number of impulses the counter has counted.

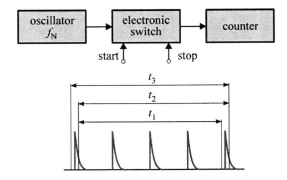

Figure 10.55 Time measurement

The frequency can be measured with the counter counting the impulses during a known time span. The impulses from the generator are connected to the electronic switch through the frequency divider.

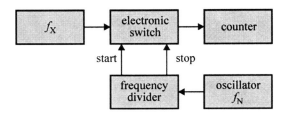

Figure 10.56 Measurement of high frequency

The first impulse opens the switch and, after a predetermined time (1 second, for example), the second impulse closes the switch. This method is appropriate for high frequency measurements (Figure 10.56).

The measurement of low frequencies can be carried out by measurement of one period (or 10, 100, 1000 periods) of sine wave. One impulse from the measured source opens the switch, and the second closes it, while the counter counts the impulses from the generator (Figure 10.57).

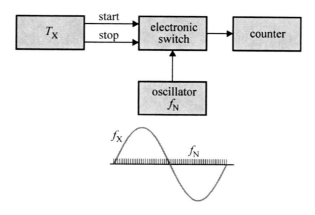

Figure 10.57 Measurement of low frequency

Selected bibliography:

1. *Insulation Resistance Measurement*, http://www.electrotechnik.net/2009/05/insulation-resistance-measurement.html, Electrotechnik, A Magazine for Electrical Engineering, Retrieved October 20th, 2010
2. *Earth Resistance Measurement*, http://www.electrotechnik.net/2009/05/earth-resistance-measurement.html, Electrotechnik, A Magazine for Electrical Engineering, Retrieved October 20th, 2010
3. V.Bego, "*Measurements in Electrical Engineering*", (in Croatian) Tehnička knjiga, Zagreb, 1990
4. D. Vujević, B.Ferković, "*The Basics of the Electrotechnical Measurements*, part I, (in Croatian), Zagreb 1994
5. D. Vujević, B.Ferković, "*The Basics of the Electrotechnical Measurements*, part II, (in Croatian), Zagreb 1996
6. D. Vujević, "*Measurements in Electrical Engineering – Laboratory Experiments*", (in Croatian), Zagreb, 2001
7. "*Calibration: Philosophy in Practice*", Fluke, Second Edition, 1994.
8. M. E. Cage, D. Yu, B. M. Jeckelmann, R. L. Steiner, R. V. Duncan: "*Investigating the Use of Multimeters to Measure Quantized Hall Resistance Standards*", IEEE Trans. Instrum. Meas. IM-40 2, pp. 262-266., 1991.
9. D. Braudaway: "*Precision Resistors: A Review of Techniques of Measurement, Advantages, Disadvantages*", IEEE Trans. Instrum. Meas. Vol. 48, No. 5, pp. 884-888, October 1999.
10. D. Braudaway: "*Behavior of Resistors and Shunts: With Today's High-Precision Measurement Capability and a Century of Materials Experience, What Can Go Wrong?*", IEEE Trans. Instrum. Meas. Vol. 48, No. 5, pp. 889-893, October 1999.

11. Electromagnetic Metrology, *DC & Low Frequency Measurement Services Booklet*, National Physics Laboratory, 1999.
12. H. L. Curtis, L. W. Hartman: "*A Dual Bridge for the Measurement of Self Inductance in Terms of Resistance and Time*", Journal of Research of NIST, pp. 1-13, July 1940.
13. A. Daire: "*Test Instruments: Improving the Repeatability of Ultra-High Resistance and Resistivity Measurements*", Keithley Instruments, http://www.keithley.com/white_papers/ultra_wp/index.html., Retrieved, October 20th, 2010
14. "*Electromagnetic Measurements*", NIST Calibration Services User Guide, pp. 114-117, 1991.
15. R. F. Dziuba: "*Automated Resistance Measurements at NIST*", Proc. 1995 National Conference of Standards Laboratories (NCSL), July 1995, pp. 189-195.
16. *Sci-Tech Encyclopedia*, McGraw-Hill Encyclopedia of Science and Technology, 5th edition
17. *Agilent Impedance Measurement Handbook*, A Guide to Measurement Technology and Techniques, 4th Edition, http://cp.literature.agilent.com/litweb/pdf/5950-3000.pdf, Retrieved, October 20th, 2010
18. All about circuits, *AC bridge circuits*, http://www.allaboutcircuits.com/vol_2/chpt_12/5.html, Retrieved, October 20th, 2010

11. INSTRUMENTATION AND COMPUTERS

11.1 HISTORY OF INSTRUMENTATION AND COMPUTERS, INTERFACES, AND BUSES

In the past, every measurement instrument required a person to operate it. As the whole measurement process was managed by people, it was a slow, tiring, and error-prone process. It was also susceptible to human errors in writing and managing the measurement results, difficult to synchronize multiple instruments, and impossible to automatize the whole measurement process.

When there is a need to simultaneously manage a large number of instruments in a measuring system, it become necessary to automatically control these instruments and to manage a large number of measurement results, sometimes in a very short period of time and in a specific order. This is usually achieved by using a specific program to automatically control the measurement process. With advancement in computers and digital instruments, it also became necessary to connect several instruments in the measuring system, which is where a computer has several important roles: managing the measurement process, calculating measurement data, and sending commands to measuring instruments (Figure 11.1). In the late 1960s, several manufacturers started developing specialized interfaces and circuits for connecting instruments and computers. Instead of controlling each of the instruments individually, the operator used the computer to control all instruments in a measurement system. The computer also handled data and measurement results. However, there was no standardized solution or standard, as the manufacturers developed their proprietary interfaces and standards. This situation was causing difficulties for engineers trying to assemble a measurement system comprised from instruments of different manufacturers.

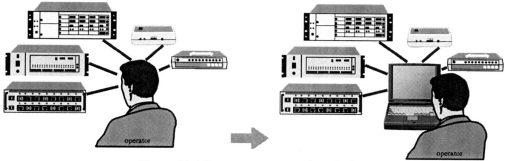

Figure 11.1 Instrument control evolution

One of these manufacturers soon provided a solution that everybody else accepted. In 1965, Hewlett Packard developed an interface called HPIB (Hewlett Packard Interface Bus) to connect computers with programmable instruments. This standard was accepted by most manufacturers and was then standardized by the international organization Institute of Electrical and Electronic Engineers (IEEE-488) in 1975. Later, other standards and interfaces for measurement systems emerged, such as USB, Ethernet, RS-232, and IEEE 1394 (FireWire), but IEEE-488 is still widely used because of its easy use and robustness. All the above mentioned standards are mostly used for standalone instruments. However, as technologies emerged for modular instrumentation such as PCI, PXI, and VXI, there was a need for standards that could exploit the advantages of such equipment, such as speed and synchronization.

11.2. INTERFACE BUSES FOR STANDALONE INSTRUMENTS

11.2.1 General Purpose Interface Bus (GPIB)

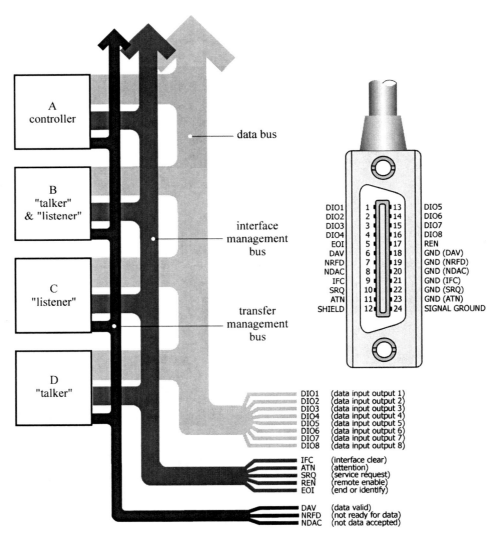

Figure 11.2 GPIB card and connector

INSTRUMENTATION AND COMPUTERS

The IEEE-488, published in 1975 and entitled "Standard Digital Interface Programmable Device," allowed devices from different manufacturers to be included in measurement systems, provided they respect the standard. The standard was expanded in 1978 and 1987. It describes the 8-bit parallel interface, known as GPIB (General Purpose Interface Bus), which has a maximum data rate of 1 MiB/s. The bus uses 16 active lines, while 8 lines are used for earthing (Figure 11.2).

The 8 lines serve for the transfer of measurement data: 3 are for synchronization and 5 are for addressing and bus control. The GPIB system instruments may have three different functions:

- controller
- talker
- listener

Figure 11.3 GPIB PC card

The "controller" is usually a computer with appropriate software support that manages the entire measurement system, sends commands to devices (e.g., determining the range of the measurement device, the number of measurements, etc.), and collects measurement data (e.g., measurement results, machine status, etc.).

The "talker" can only send data, and the "listener" can only receive data or commands. Some devices can be "talkers" and "listeners" at the same time, i.e., they can send and receive data (e.g. multimeter). Each device in the system has its own address (two-digit number) and the maximum number of devices that can be simultaneously connected to the system is 15. Devices can be connected in star, in line, or in both ways, while the maximum total cable length should not exceed 20 meters. Computers are usually not equipped with the GPIB card, so it must be installed later (Figure 11.3). Instruments in the system can be connected in Daisy chain (Figure 11.4) or Star bus configuration (Figure 11.5).

11.2.2 IEEE 488.2 Standard

The original standard IEEE-488 of 1975, which defined the physical, electrical, and mechanical properties of GPIB cards and interfaces was later designated as IEEE-488.1 and then in 1987, the IEEE-488.2 standard was adopted for defining commands, structure of data, and syntax. This standard also defined bus handling and setting the measurement parameters. Today, there are several million instruments that provide support for the IEEE-488 standard, as it is a proven and robust standard that has been in service for more than 30 years. It is easy to use and easy to implement for developers. Meanwhile, several other standards have been developed that enable connecting computers and measuring devices (i.e. USB and Firewire), but it is anticipated that the GPIB protocol will be used for many more years due to its good properties and high utility distribution.

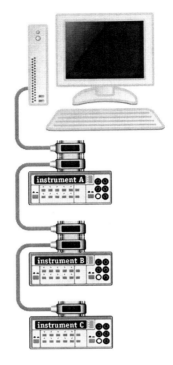

Figure 11.4 Daisy chain GPIB configuration

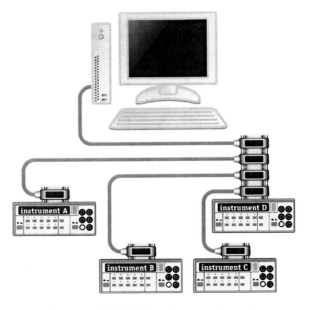

Figure 11.5 GPIB Star bus configuration

INSTRUMENTATION AND COMPUTERS

As the GPIB protocol is the most common in instruments, sometimes it is necessary to connect them to computers that do not have a GPIB interface (such as laptops). Then, so-called bridges are used (i.e., USB-GPIB) that can be used to connect different communication buses with each other (Figure 11.6).

Figure 11.6 GPIB-USB bridge

11.2.3 HS488

In 2003, the protocol HS488, developed by National Instruments, was included in the IEEE-488.2 standard, which improved the GPIB system performance, and increased the transfer rates from 1 Mbyte/s to 8 Mbyte/s. The transfer rate is actually not defined by the standard. It is not necessary that the data transfer be synchronized, as the transfer rate is handled by the software and instrument capabilities. Between two devices and 2 m of cable, the HS488 can transfer data of up to 8 Mbytes/s. For a fully loaded system with 15 devices and 15 m of cable, HS488 transfer rates can reach 1.5 Mbytes/s.

11.2.4 RS-232 and RS-485 Serial Connection

RS-232 (Recommended Standard 232) is an old standard introduced to computer technology in 1962 that implements the standard for serial binary single-ended data and control signals. It was used in computers with serial ports to connect peripheral devices such as mouses. This standard was also implemented in some test and measurement equipment. It defined the electrical characteristics and timing of signals, the meaning of signals, and the physical size and pinout of connectors. The data transfer rate was slow, up to 28.8 kB/s. This was improved over the years by introducing several new standards. The most important one was RS-485, which introduced the point-to-point, multi-dropped, and multi-point configuration. Today, the protocol can handle a maximum of 32 devices; the data transfer has also been improved over RS-232. The RS-485 connector with pin description is shown in Figure 11.7.

INSTRUMENTATION AND MEASUREMENT IN ELECTRICAL ENGINEERING

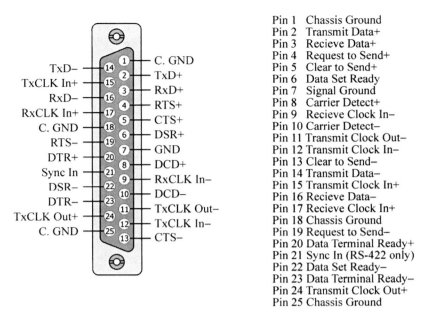

Figure 11.7 RS-485 DB25 Pinout

11.2.5 Ethernet and VXI-11 Standard

Ethernet is the name for a set of standards that provide networking in the Local Area Network (LAN), and which are described in the IEEE 802.3 standards group. Support for Ethernet is now common in all personal computers, and it is sometimes used in test and measurement equipment. The highest transmission speed increases every few years; currently, 1 Gib/s is the maximum speed for transmission over copper lines, and 10 Gib/s over the optical cable. Since the computer network based on Ethernet is extremely widespread, it is ideal for the connection and management of test and measurement equipment. The standard VXI-11 was developed to emulate interface messages of the IEEE-488 bus on the Ethernet link. Based on this standard, messages are exchanged between the controller and the instrument. The messages are identical with SCPI commands. They can be organized into four groups: program messages (control command to the instrument), response messages (values returned by the instrument), service requests (spontaneous queries of the instrument), and low–level control messages (interface messages). The number of controllers that can address an instrument is practically unlimited in the network. In the instrument, the individual controllers are clearly distinguished. This distinction continues up to the application level in the controller, i.e., two applications on a computer are identified by the instrument as two different controllers.

11.2.6 LXI – LAN Extensions for Instrumentation

LXI - LAN Extensions for Instrumentation is a relatively new protocol and is directly linked to the widely accepted protocols, which are common on local computer networks. It is designed primarily for interconnection and management of measuring instruments in the fields of aviation, medical, automotive, and defense technology. The standard addresses small, modular instruments and, except for incoming and outgoing connections, there are no control panels or indicators. They are controlled exclusively via a computer network based on the Ethernet protocol group. The great advantage of this is that we can avoid placing the control computers in the vicinity of the instruments themselves, which can be placed in another location, where only the network connection is sufficient to control the instruments. This protocol also supports synthetic instruments that perform specific tasks, such as data analysis or signal conditioning.

11.2.7 Universal Serial Bus (USB) and USTMC Class

Universal Serial Bus (USB) is a fast serial bus that is available in all personal computers, but increasingly in measuring equipment as well. There are several versions of this protocol with different transmission speeds:

- USB 1.1 (September 1996) - Allows transmission data rate from 1.5 Mib/s to 12 MiB/s
- USB 2.0 (April 2000) - Enables transmission data rate from 480 Mib/s to 60 MiB/s
- USB 3.0 - Currently still in the development stage, but with the application of optical fiber, will enable a transmission data rate of up to 600 MiB/s.

The cables that are used in versions 1.1 and 2.0 are made of unshielded copper wires; they are not suitable for all applications because of their high sensitivity to external electromagnetic interference. An additional drawback is the maximum cable length of five meters in the USB 2.0 standard, which significantly complicates the protocol's use in laboratory and industrial environments. This deficiency can be avoided by using accessories to extend the range to 50 m (using four copper pairs) and to even 10 km (using optical lines).

USBTMC (USB Test and Measurement Class) is a protocol that upgrades the USB protocol and makes it possible to communicate with instruments similar to the GPIB protocol. Seen from the user's side, the USB device behaves in the same way as the GPIB device, as it supports all the features of that protocol. In this way, manufacturers of measuring equipment can easily upgrade the communications interface and still maintain software compatibility.

11.2.8 IEEE 1394

Developed by Apple Inc. in the late 1980s and known under the brand names of companies that use it, such as FireWire (Apple), i.Link (Sony) or DV (Panasonic), IEEE-1394 was initially designed as a replacement for the parallel SCSI (Small Computer System Interface) bus, but finds its widest application in amateur digital audio and video equipment. It is also used in the control of industrial robots, the car industry (IDB-1394 Customer Convenience Port), and the aerospace (IEEE 1394b) industry.

INSTRUMENTATION AND MEASUREMENT IN ELECTRICAL ENGINEERING

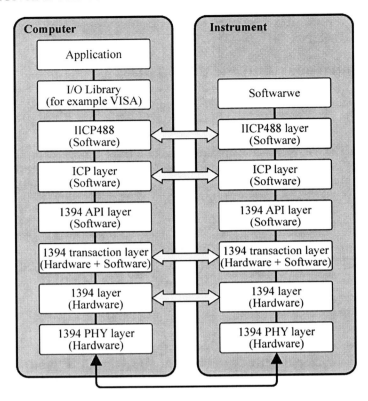

Figure 11.8 IICP488 communication model

The data bus can handle transfer speeds of up to 800 Mib/s, and in the future, it may be possible to use fiber connections that will achieve transmission speeds to 6.4 Gib/s. Industrial instrumentation control protocol, **IICP**, is a test and measurement protocol being developed as the extension of 1394 asynchronous communication to test and measurement applications (Figure 11.8). The IICP488 layer implements the services for GPIB-like communications, such as sending a message, sending a device command, or setting local operation.

11.2.9 Comparison of Interface Buses for Standalone Instruments

Each of the presented buses has its own comparative advantages over other solutions, and in Table 11.1 the comparison between the different protocols is clearly presented.

Table 11.1 Comparison of different protocols

	GPIB	RS-232	Ethernet	USB	IEEE 1394
Maximum speed	1.5 Mb/s 8 Mb/s (HS488)	28.8 kB/s	125 Mb/s	60 Mb/s	100 Mb/s
Availability	very large	large	small	small	very small
Delay (First Byte and Service Request)	small	small/medium	large	large	large
Maximum number of devices	14	1	No limit	127	63
Maximum length of cable	20 m	15 m	no limit (150 m wireless)	30 m	4.5 m

INSTRUMENTATION AND COMPUTERS

The modern oscilloscope back portion of ports can be seen in Figure 11.9.

Figure 11.9 Modern oscilloscope with LXI (LAN), USB, GPIB, and video connectivity

11.3. INTERFACE BUSES FOR MODULAR INSTRUMENTS

In recent times, modular instruments are replacing standalone instruments in certain applications. Modular instruments have several advantages, as they can be assembled from multi-vendor components, making them user-defined, rather than vendor-defined. Also, modular instruments are usually easily integrated into the final product, and can easily be upgraded and expanded to add new functionalities. In a way, it is a "pay for what you need," rather than "pay for what you get" instrument. Also, the measurement system assembled from modular instrumentation has increased measurement throughput and enabled faster data transfer, better instrument synchronization, and tighter software and hardware integration, resulting in a lower cost and smaller footprint.

11.3.1 Peripheral Component Interconnect (PCI) and PCI Express Bus

The PCI bus is an industry standard in computer architecture that has robust electrical specification and flexible mechanical specification. It is one of the most commonly used internal computer buses today. With a shared bandwidth of 132 MB/s, PCI offers high-speed data streaming and deterministic data transfer for single-point control applications. There are many different data acquisition hardware options for PCI, with multifunction I/O boards of up to 10 MS/s and up to 18-bit resolution. PCI Express, an evolution of PCI, has dedicated bus bandwidth provided by independent data transfer lines. Unlike PCI, in which 132 MB/s of bandwidth is shared among all devices, PCI Express uses independent data lanes that are each capable of data transfer of up to 250 MB/s. The PCI Express bus is also scalable from a single x1 (pronounced "by one") data lane to x16 data lanes for a maximum throughput of 4 GB/s, enough to fill a 200 GB hard drive in less than a minute. For measurement applications, this means higher-sustained sampling and data rates because multiple devices do not have to compete for time on the bus. Test and measurement equipment that uses the PCI bus is sometimes called a **computer instrument**, as it can be installed on a computer like any other PCI card.

11.3.2 VXI (VMEbus eXtensions for Instrumentation) Bus

The VXI bus is based on the Versa Module Eurocard (VME) backplane. It was standardized by the IEEE 1155 standard in 1993. It was the first attempt to create an open multi-vendor system that was supported by many vendors all over the world. The VXI bus offers industry-standard instrument-on-a-card architecture with rack-and-stack instrumentation that resulted from physical size reduction. Also, the system offers tighter timing and synchronization between multiple instruments, and faster transfer rates than the 1 Mb/s rate of the 8-bit GPIB. In the original VME bus, the 40 Mb/s data rate was standard; in VME-64 this was improved to 80 Mb/s. Although VXI systems today can be purely VXI, it is also possible to integrate the VXI system with other available systems consisting of GPIB instruments, VME cards, or plug-in data acquisition (DAQ) boards. Even though VXI had some success, such as physical size reduction, more precise timing, and interchangeability between any vendors, it also had some shortcomings, as there was no software standard, no greatly reduced system cost, and the ease of use was not achieved as planned. The typical VXI system is shown in Figure 11.8.

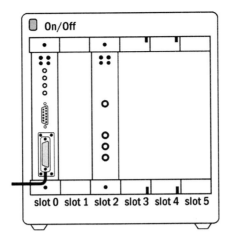

Figure 11.8 VXI chassis

11.3.3 PCI eXtensions for Instrumentation (PXI)

PCI eXtensions for Instrumentation (PXI) were developed to bridge the gap between desktop PC systems and high-end VXI and GPIB systems. PXI is a rugged PC-based platform that offers a high-performance, low-cost deployment solution for measurement and automation systems. PXI combines the Peripheral Component Interconnect (PCI) electrical bus with the rugged, modular Eurocard mechanical packaging of the Compact PCI, and adds specialized synchronization buses and key software features. The PXI system (Figure 12.3.) uses the computer as the **control unit** (**controller**), but the whole system (control unit and instruments) is enclosed in one case (**chassis**), and **modular instruments** are connected with an inexpensive and fast PCI bus. The PXI chassis and controller with one module can be seen in Figure 11.9.

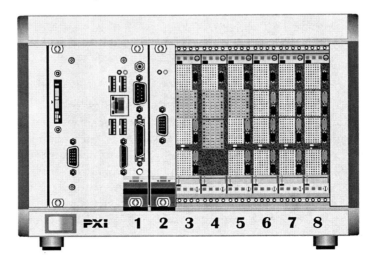

Figure 11.9 PXI chassis with controller board and expansion slots

In this way, 32-bit and 64-bit data exchanges with a maximum data transfer rate of 264 Mb/s are achieved. The PXI system also allows synchronization between the various modular units and the control unit with the internal clock of 10 MHz. PXI Timing and Triggering Buses can be seen in Figure 11.10.

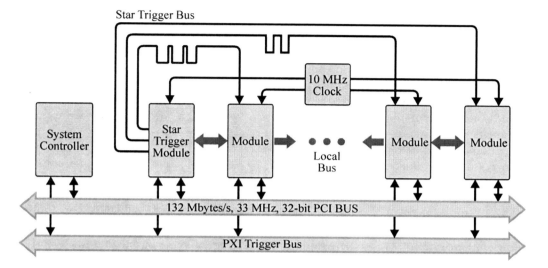

Figure 11.10 PXI timing and triggering

All PXI systems must have drivers for the Windows operating system, and thereby enable a wide selection of software. As the computer monitor is used as the user interface, the user himself will determine the front panel of the instrument. Such instruments are called **virtual instruments**. Software support is an essential part of such systems. Although such systems can be programmed using **text** languages such as **Pascal** or **C**, today there are several software systems for the **graphic programming** of virtual instruments; among them, the de-facto standard software system is LabVIEW from National Instruments. The PXI system had success where VXI didn't, as it is based on the widely used PCI bus, and today it is the most used standard for modular instrumentation in the world, featuring precise synchronization, software support, and integrated hardware triggering.

11.3.4 Wireless Connectivity in Measurement Applications

Wireless technology extends the flexibility and portability of measurement applications where cables are inconvenient or impractical. However, wireless technology also has the highest latency of any other data acquisition bus, so this technology cannot be used where high speed is required. There are many different wireless technologies available, the most popular of which is IEEE 802.11 (Wi-Fi). For smaller data rate applications, such as sensor networks, Bluetooth or ZigBee wireless technologies are also used.

11.4 SOFTWARE SUPPORT FOR INSTRUMENT CONTROL

The rapid development of personal computers in recent decades has led to a revolution in the field of testing equipment, measurement, and automation. The advancement in hardware standards and protocols logically had to be followed by software tools, applications, and drivers to ease the development of testing and measurement methods and procedures. In the beginning, each manufacturer developed their own language for the control of their instruments. Even different instruments from the same manufacturer had different software protocols. As the instruments became more complex, so did their control systems. Engineers had to build a measurement system while also learning different languages and syntaxes for different instruments, which was very time-consuming.

11.4.1 Virtual Instrumentation

One output of the spread of personal computers in the field of metrology is virtual instrumentation technology. A virtual instrument consists of a PC or workstation equipped with the appropriate software and hardware (both standalone and modular instruments), which together can perform all operations of classical measuring instruments. The operator can also use the virtual instrument run on a remote computer. This idea represents a significant shift away from a purely hardware-oriented instrument, as the computational potential of personal computers, as well as their ability to connect with the outside world, allows them to be used to the their full potential. The components of virtual instrumentation can also be reused without additional investment, which makes them economical and flexible; in contrast, proprietary instruments which have the control panel and internal electronics predetermined have a low possibility to add functionality as the need arises. Virtual instruments installed on personal computers are able to take advantage of the latest technologies of both the hardware and software that are built into them. It is also possible to use portable computers to make a mobile measurement system. The use of virtual instrumentation, therefore, cuts on development, production, and maintenance costs. An important development in this field is the expansion of computing technology to the embedded systems where whole measurement systems are embedded on a single chip with multiple connectivity options.

11.4.2 Graphical Programming

Although virtual instruments can be programmed using text-based languages, such as C++ or LabWindows, graphical programming languages such as the National Instruments' LabVIEW or Agilent's VEE are easier to use, especially for beginners. With this type of language, people without any programming knowledge or experience can create complex virtual instruments relatively quickly. Graphical programming is also sometimes called data flow programming, as it follows a data path between certain operations.

11.4.3 LabVIEW™ Graphical Programming Language

LabVIEW™ is a graphical programming language that uses data flow to control the program execution. Unlike text-based program languages, LabVIEW™ uses icons and lines. In C language, the FOR loop program would be written as:

```
scanf ("%d", &num_of_elem);
sum = 0;
for (i=0; i<num_of_elem; i++)
    sum += i + 1;
printf ("%d", sum);
```

The "FOR loop" written in LabVIEW™ is shown in Figure 11.11.

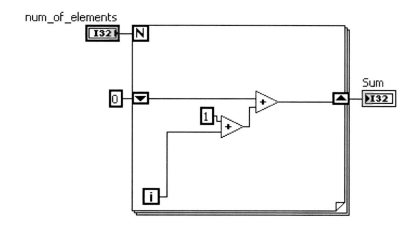

Figure 11.11 "FOR loop" in LabVIEW™

Although it can be used to develop general purpose applications, the LabVIEW™ programming language is primarily intended for the development of user-defined virtual instrumentation, automation, measurement systems, monitoring, and control applications. It also features advanced analysis of measurement data in real time. LabVIEW™ has rapid application development support for a large number of modern hardware technologies such as GPIB, VXI, PXI, RS-232, RS-485, and modular instruments. LabVIEW™ is also used for the development of remote laboratories using its integrated Web server. LabVIEW™ programs are called virtual instruments (VI). Like all other development environments, LabVIEW™ programs can compile and create an executable (EXE) file or shared library (dll). The virtual instrument consists of two main components: the control panel (front panel) and block diagrams (block diagram). There are also other components, such as the icon editor and connector pane.

The **front panel** is a user interface that contains controls and indicators (inputs and outputs). The controls and indicators can be numerical or graphical, and since there are many available in the LabVIEW™ library, the user can create a virtual instrument easily.

Once the front panel is assembled out of all the necessary controls and indicators, the controls and indicators must be connected in a block diagram. The **block diagram** is a programming code, and all controls and indicators from the front panel are displayed in a block diagram as a terminal like this: DBL. All objects deleted or added to the block diagram or front panel are automatically added in the other window.

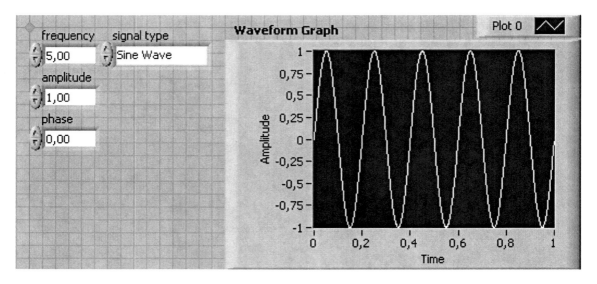

Figure 11.12 Front panel for sine wave generator program written in LabVIEW™

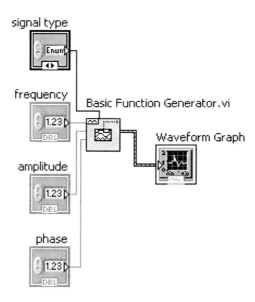

Figure 11.13 block diagram for sine wave generator written in LabVIEW™

Objects in the block diagram include terminals from front panel objects, but also objects that are not displayed on the front panel, such as functions, subroutines (subVI), constants, structures, and wires that connect the various objects in the block diagram to make it a functional program. The following example shows the front panel (Figure 11.12) and the corresponding block diagram (Figure 11.13) for a simple sine wave generator program.

INSTRUMENTATION AND COMPUTERS

The program flow is presented with lines and graphical symbols. In both the front panel and block diagram, in the upper right corner there is an icon 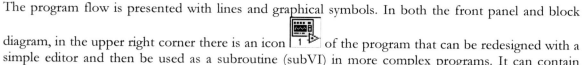 of the program that can be redesigned with a simple editor and then be used as a subroutine (subVI) in more complex programs. It can contain simple pictures and/or text. Also, a proper set of inputs and outputs must be defined in a similar fashion, like defining the parameters of procedures and functions in C or Pascal. If the icon is pressed by the right mouse button, a Connector pane appears. LabVIEW™ is designed in such a way that the programs can be used like all other functions already built into LabVIEW, so the programmer can build programs in a hierarchical way.

11.4.4 IEEE 488.2

In 1987, the IEEE-488.2 standard was the first standard that defined commands, the structure of data, and syntax. It also defined bus handling, status byte, and setting measurement parameters etc., but only for IEEE-488 communication through the GPIB interface bus. The meaning of each bit in the status byte was standardized. The IEEE 488.2 mandatory control sequence is shown in Table 11.2. IEEE 488.2 defined 15 required and four optional control sequences that specify the exact IEEE 488.1 messages that are sent from the controller, as well as the ordering of multiple messages.

Table 11.2 IEEE 488.2 Mandatory Controller Sequences

Description	Control Sequence
Send ATN-true commands	SEND COMMAND
Set address to send data	SEND SETUP
Send ATN-false data	SEND DATA BYTES
Send a program message	SEND
Set address to receive data	RECEIVE SETUP
Receive ATN-false data	RECEIVE RESPONSE MESSAGE
Receive a response message	RECEIVE
Pulse IFC line	SEND IFC
Place devices in DCAS	DEVICE CLEAR
Place devices in local state	ENABLE LOCAL CONTROLS
Place devices in remote state	ENABLE REMOTE
Place dev. in remote with local lockout state	SET RWLS
Place devices in local lockout state	SEND LLO
Read IEEE 488.1 status byte	READ STATUS BYTE
Send group execution trigger (GET) message	TRIGGER

The IEEE 488.2 Common commands and queries are shown in Table 11.3. Since these operations are common to all instruments, IEEE 488.2 defined the programming commands used to execute these operations and the queries used to receive common status information.

The IEEE 488.2 controller protocols are shown in Table 11.4. Protocols are high-level routines that combine a number of control sequences to perform common test system operations. IEEE 488.2 defines two required and six optional protocols. These protocols merge several commands to execute the most common operations. The mandatory protocol RESET ensures that the GPIB has been initialized and all devices have been cleared and set to a known state. The ALLSPOLL protocol serial polls each device and returns the status byte of each device.

Table 11.3 IEEE 488.2 Common commands and queries

Mnemonic	Group	Description
*IDN?	System Data	Identification query
*RST	Internal Operations	Reset
*TST?	Internal Operations	Self-test query
*OPC	Synchronization	Operation complete
*OPC?	Synchronization	Operation complete query
*WAI	Synchronization	Wait to complete
*CLS	Status and Event	Clear status
*ESE	Status and Event	Event status enable
*ESE?	Status and Event	Event status enable query
*ESR?	Status and Event	Event status register query
*SRE	Status and Event	Service request enable
*SRE?	Status and Event	Service request enable query
*STB?	Status and Event	Read status byte query

Table 11.4 IEEE 488.2 controller protocols

Keyword	Name	Compliance
RESET	Reset System	Mandatory
FINDRQS	Find Device Requesting Service	Optional
ALLSPOLL	Serial Poll All Devices	Mandatory
PASSCTL	Pass Control	Optional
REQUESTCTL	Request Control	Optional
FINDLSTN	Find Listeners	Optional
SETADD	Set Address	Optional, but requires FINDLSTN
TESTSYS	Self-Test System	Optional

11.4.5 Standard Commands for Programmable Instruments

Later, in 1990, the Standard Commands for Programmable Instruments (SCPI) was introduced, defining a standard for syntax and commands to use in control of programmable test and measurement devices. It was aimed to ease the development and maintenance of measurement systems, as prior to this standard, each manufacturer developed their own command sets for programmable instruments. This lack of standardization forced test system developers to learn a number of different command sets and instrument-specific parameters for the various instruments used in an application, leading to programming complexities and resulting in unpredictable schedule delays and development costs. While originally developed for IEEE-488, SCPI can also be used with other standards. It requires conformance to IEEE-488.2, but is a pure software standard. SCPI syntax is ASCII text, and therefore can be attached to any computer test language, which makes it hardware independent. As a means of achieving compatibility and categorizing command groups, SCPI defined a model of a programmable instrument. The structure hiararchy of SCPI is shown in Figure 11.14.

INSTRUMENTATION AND COMPUTERS

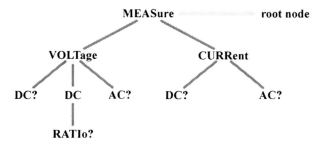

Figure 11.14 SCPI command structure

The following command instructs a digital multimeter to make a voltage measurement on a 10 V range with a 0,001 V resolution.

MEASure:VOLTage:DC? 10, 0.001

11.4.6 Virtual Instrument Software Architecture (VISA)

Virtual Instrument Software Architecture (VISA) is an industry standard for communication of instruments and computers, providing the programming interface between the hardware and development environments. It is widely used by many manufacturers of measurement instruments. It includes specifications for communication with resources that can be instruments, but may be other resources. VISA provides the interchangeability and reuse of software from different manufacturers and for many different programming languages. It can also provide configuration and programming of many systems and protocols such as GPIB, VXI, PXI, Serial, Ethernet, and USB. The most important objects in VISA are **resources**. The functions that can be used with an object are operations. The object also has variables or attributes that contain information related to the object. A resource can be any instrument in the system, including serial and parallel ports. To communicate with the resource, a **session** must be opened, and a VISA session number is given, which is a unique handle to that instrument. This session number is used in all subsequent VISA functions. In LabVIEW, the session number is called "refnum." Resources are called "instrument descriptor." This term specifies the interface type (GPIB, VXI, ASRL), the address of the device (logical address or primary address), and the VISA Session type (INSTR or Event). The following list gives the instrument syntax for the serial, GPIB, and VXI instrument types:

    Serial   ASRL[number][::INSTR]

    GPIB   GPIB[number]::address[::INSTR]

    VXI   VXI[number]::VXI address[::INSTR]

VISA can be considered an application programming interface (API) between the hardware and bus of the measurement system and programming tools such as LabVIEW™ or C. The simple VISA program written in LabVIEW™ can be seen in Figure 11.15. Both examples use NI-VISA, the National Instruments implementation of the VISA standard. The VISA Write block writes the data buffer specified to the instrument indicated by the VISA session or resource name. Then, the VISA Read function reads back 1024 bytes from the instrument.

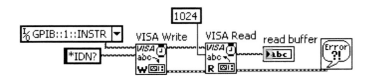

Figure 11.15 Simple VISA program in LabVIEW

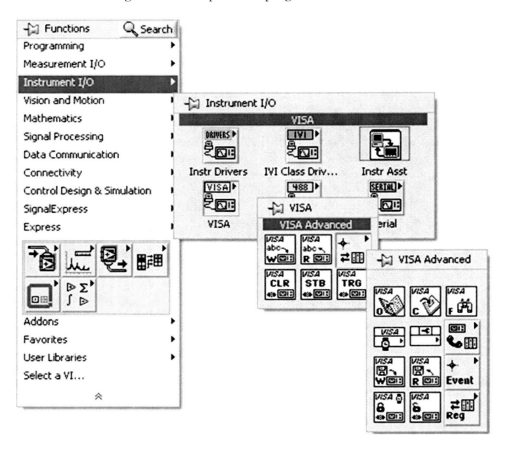

Figure 11.16 Software control panel for VISA in LabVIEW

The following example uses NI-VISA in National Instruments LabWindows/CVI, a development environment built on the ANSI C programming language. The program has the same functionality but includes session management functions handled automatically in LabVIEW such as VISA Open and VISA Close.

    ViSession drm, io;
    ViUInt32 count;
    ViChar buf[1024];
    viOpenDefaultRM (&drm);
    viOpen (drm, "GPIB::1::INSTR", 0, 0, &io);

```
viWrite (io, "*IDN?\n", 6, &count);
viRead (io, buf, 1024, &count);
printf ("Identity: %*s\n", count, buf);
viClose (drm);
```

Both examples include a VISA session resource descriptor. It is a GPIB instrument at address 1 with the descriptor GPIB::1::INSTR.

In graphical programming languages such as LabVIEW, the NI-VISA software control panel is available for programmers including many other functions (Figure 11.16).

11.4.7 Instrument Drivers

An instrument driver is a set of software routines that control a programmable instrument. Each routine corresponds to a programmatic operation such as configuring, reading from, writing to, and triggering the instrument. Instrument drivers simplify the instrument control and reduce the test program development time by eliminating the need to learn the programming protocol for each instrument. In the past, every manufacturer designed drivers for their own equipment with no standardized procedures. This was changed by IVI foundation whose goal was to:
- make drivers hardware-interchangeable,
- allow users to exchange instruments in their system with little or no software modifications,
- improve driver quality by defining specifications for drivers to ensure more consistent quality and robustness,
- provide software interoperability by defining an architecture to easily integrate software from multiple vendors,
- and provide a consistent and standard method for access to driver capabilities.

The IVI compatible driver consists of two layers: the specific and the class layer. One class driver is used for each class of instruments, so it is possible to create a driver and application that do not contain any reference to a specific instrument and do not use that instrument's driver API. The specific driver contains functions specific to a certain instrument. To enable interchangeability, the foundation creates IVI class specifications that define the base class capabilities and class extension capabilities for some of the most popular instrument classes. There are currently eight instrument classes defined:

- DC power supply
- Digital multimeter (DMM)
- Function generator
- Oscilloscope
- Power meter
- RF signal generator
- Spectrum analyzer
- Switch

Another feature of IVI is that not every feature of every instrument has to fit into the generic classes. Instruments have unique features that do not fit into classes. If there is a need for some specific feature, then there is a specific driver that will have the needed functionality. Therefore, the user can call a specific driver, a class driver, or mix them (Figure 11.17).

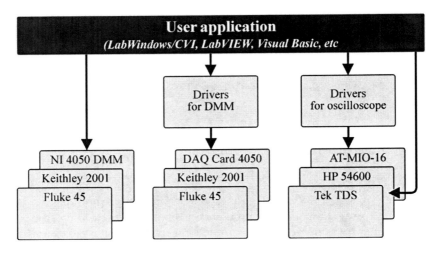

Figure 11.17 IVI drivers from the programmer's view

From the user's perspective, you can call a specific driver directly, call a generic driver to get interchangeability, or mix them to take advantage of specific features. The LabVIEWTM IVI driver control panel can be seen in Figure 11.18.

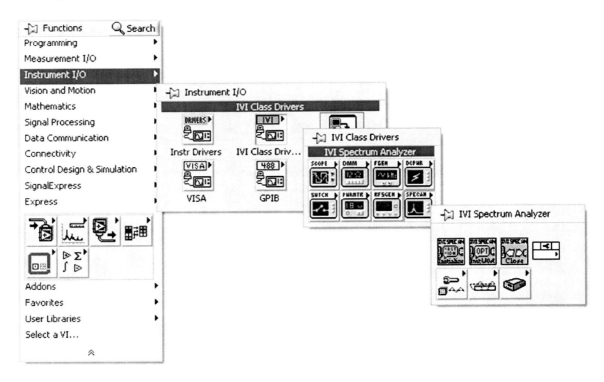

Figure 11.18 IVI driver control panel in LabVIEW

A simple program for instrument control using IVI drivers is seen in Figure 11.19. This VI uses class drivers for the digital multimeter and performs a single immediate triggering to read a single measurement from the instrument. The user can easily change the multimeter parameters and as the IVI is a class driver, it should work with any multimeter that supports the simple DC Volts measurement function.

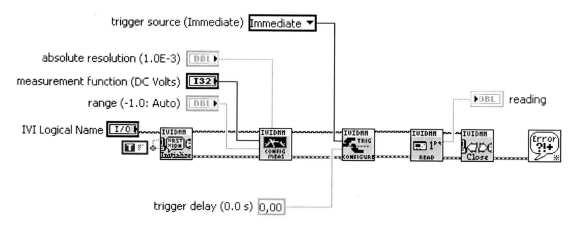

Figure 11.19 Simple program using IVI drivers to measure DC voltage

Selected bibliography:
1. *The HS488 Protocol*, http://zone.ni.com/devzone/cda/tut/p/id/4283, Retrieved October 16th, 2010
2. *SCPI Consortium*, http://www.ivifoundation.org/scpi/default.aspx, Retrieved October 16th, 2010
3. Dan Mondrik, *Advanced Software Techniques for Instrument Control*, National Instruments Week, August 2000
4. *RS-232 Standard*, http://en.wikipedia.org/wiki/RS-232, Retrieved October 16th, 2010
5. Marko Jurčević, *System for Remote Calibration and Testing of Measurement Equipment*, doctoral dissertation, Zagreb, 2010
6. *VXI-11 Interface*, http://www.rohde-schwarz.com/webhelp/zvb/HW_Interfaces/LAN_Interfaces/VXI-11_Interface.htm, Retrieved October 16th, 2010
7. *An Introduction to the Instrument and Industrial Control Protocol*, http://www.1394ta.org/press/WhitePapers/iicpPaper2.pdf, Retrieved October 16th, 2010
8. *Choosing the Right Bus for Your Measurement Application*, http://zone.ni.com/devzone/cda/tut/p/id/9401, Retrieved October 16th, 2010
9. *Short Tutorial on VXI*, http://zone.ni.com/devzone/cda/tut/p/id/2899, Retrieved October 16th, 2010
10. *What is an Instrument Driver*, http://zone.ni.com/devzone/cda/tut/p/id/4803, Retrieved October 16th, 2010
11. *IviDmm, Single Point Measurement*, NI Example finder, LabVIEW 8.6 version

12. MEASUREMENT SYSTEMS

12.1 MEASUREMENT SYSTEMS OVERVIEW

Measurement systems are defined as any of the systems used to associate numbers with physical quantities and phenomena, such as temperature, pressure, velocity, flow, and many others. After the physical phenomenon is converted to numbers, it can be analyzed and presented. Sometimes a measurement system is referred to as a data acquisition system. Measurement systems range from large distributed measurement systems containing numerous instruments and components to measurement systems that can be placed on a single chip. However, all these measurement systems incorporate similar building blocks (Figure 12.1) which include: sensors, signal conditioning, analogue to digital conversions, and application software. Analogue to digital converters were partly covered in Chapter 9, and the application software was mentioned in Chapter 11, so in this chapter, sensors and signal conditioning will be covered in more detail.

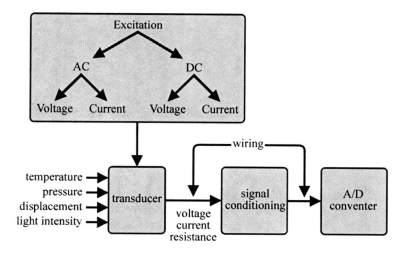

Figure 12.1 Measurement system

12.2 SENSORS AND TRANSDUCERS

A sensor is a device that converts a physical quantity (temperature, pressure, displacement, and many others) into a **signal** (usually voltage or a current), which can be then read by an instrument or data acquisition system. The term "transducer," sometimes used as a synonym for a sensor, is in fact a broader term, describing a device that transforms one type of energy to another. For example, actuators - such as electric motors - are transducers, as they transform one type of energy into another. There are two types of sensors: the self-generating type that does not require external power, and the passive type that requires external power, as the output is a measure of some variation (resistance, capacitance, etc.).

In the following chapters, some of the most popular sensors and transducers will be described.

12.2.1 Thermocouples

The thermocouple is the most popular sensor for measuring temperature and has a wide range of applications. Its work is based on the Seeback or thermoelectric effect which implies that any conductor subjected to a thermal gradient will generate a voltage. As different metals produce different voltages, the sensor is built of two dissimilar metals, so that the difference of the generated voltages can be used for measurement purposes. That difference increases with temperature and is in the range of μV/°C. However, as thermocouples are connected to the DAQ device, another junction is created (called a "cold or reference junction", see Figure 12.2), which in itself acts as a thermocouple. Thermal EMF can be calculated as:

$$E = c(T_1 - T_2) + k(T_1^2 - T_2^2) \qquad [12.1]$$

where c and k are thermocouple material constants, and T_1 and T_2 are temperatures of hot and cold junctions.

The voltage produced by the cold junction must be removed from the measurement. This is done with a procedure called "cold junction compensation" that can be realized by a special circuitry utilizing an appropriate voltage to cancel the cold junction voltage, or by software calculation using look-up tables or polynomials.

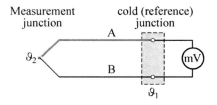

12.2 Principle of thermocouple sensor

There are several popular thermocouple types in the market:

- Type K : Chromel-Alumel
- Type J : Iron-Constantan
- Type E : Chromel-Constantan
- Type N: Nicros-Nisil
- Type T : Copper-Constantan
- Type C : Tungsten-Rhenium

Types B, R, and S use Platinum–Rhodium alloys of different percentages for each conductor. Different characteristics of output voltage and temperature can be seen in Figure 12.3.

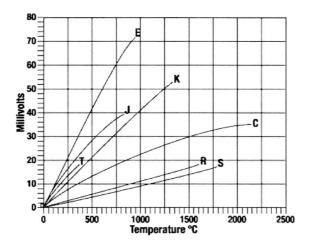

Figure 12.3 Voltage temperature relationships for different thermocouple types

12.2.2 Resistance Thermal Devices (RTDs)

Resistance thermal devices are temperature sensors that consist of a wire coil or deposited film of pure metal, whose resistance changes predictably with temperature. Metals with a high temperature coefficient can be used for this purpose (nickel, tungsten, copper), but most of the commercial RTDs are made of platinum, and are sometimes referred to as platinum resistance thermometers (PRTD). The resistance values range from 10 Ω to several thousand ohms for metal type RTDs. The mostly used platinum type RTD is Pt 100 (Figure 12.3b), with a resistance of 100 Ω at 0 °C. Platinum resistance thermometers are used for the measurement of temperature from -200 °C to 700 °C (Figure 12.4).

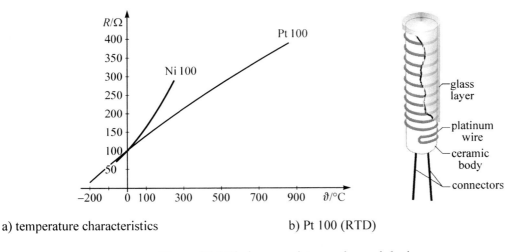

a) temperature characteristics   b) Pt 100 (RTD)

Figure 12.4 Platinum resistance thermal device

RTDs are passive sensors that require current to flow through the RTDs to produce voltage that can be measured by a DAQ device. As the resistance of RTD is low and changes slowly with temperature (usually 100 Ω, while 10 Ω RTDs are also available), special care must be taken to minimize the errors from lead resistance. This can be achieved by using the Wheatstone bridge with a 3 wire connector arrangement (Figure 12.5).

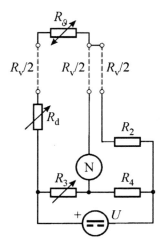

Figure 12.5 RTD in bridge connection

The bridge consists of three resistors that have a zero temperature coefficient. The RTD is separated from the bridge, so that temperature changes will only affect the RTD, and not the bridge itself. However, this separation must be accomplished with a pair of extension wires. These two wires must be the same length so their resistance effects R_L cancel each other out, as they are in opposite legs of the bridge. The third wire leads no current and acts as a sense. The output voltage is not directly proportional to the temperature, so there must be some additional calculation. The resistance R_g of the RTD in a 3-way bridge arrangement can be calculated as:

$$R_g = R_3 \left(\frac{V_S - 2V_0}{V_S + 2V_0} \right) - R_L \left(\frac{4V_0}{V_S + 2V_0} \right) \qquad [12.1]$$

Therefore, the lead resistance R_L will not add to measurement errors only if the bridge is in balance (e.g. $V_0=0$). To increase the accuracy and reliability of the resistance measurement, a 4-wire bridge connection can be used as well. However, the 4-wire resistance measurement can also be done using the 4-wire Ohms measurement method with current source and digital voltmeter (Figure 3.6).

Even though RTDs have more linear temperature-resistance dependence than the thermocouple, curve fitting is still required. The Callendar-Van Dusen equation is used to approximate the RTD curve:

$$R_T = R_0 + R_0 \alpha \left[T - \delta - \left(\frac{T}{100} - 1 \right) \left(\frac{T}{100} \right) - \beta \left(\frac{T}{100} - 1 \right) \left(\frac{T}{100} \right)^3 \right] \qquad [12.2]$$

where R_T is the resistance at temperature T, R_0 the resistance at temperature 0, and α, β and δ the temperature coefficients of a particular RTD. Polynomials have also been used for this purpose since 1968.

MEASUREMENT SYSTEMS

12.2.3 Thermistors

Thermistors are types of resistors whose resistance changes significantly with temperature (Figure 12.6).

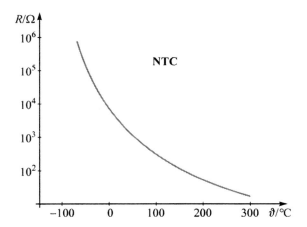

Figure 12.6 Temperature-resistance change of thermistors

Unlike RTDs, which are made of metals, thermistors are usually ceramic or polymer. Also, their temperature coefficient is usually negative and non-linear unlike RTDs, which have a positive temperature coefficient and more linear characteristics. They usually have smaller dimensions and larger error limits than RTDs (Figure 12.7).

Figure 12.7 Types of thermistors

12.2.4 Strain Gauges

Strain is the fractional change in length or deformation of a body caused by an applied force. Strain can occur both perpendicular or parallel to the applied force.

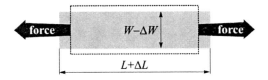

Figure 12.8 Strain of the object

Perpendicular strain is called transverse strain and is calculated as $\varepsilon = \dfrac{\Delta L}{L}$. Parallel strain is called principle strain and is calculated as $\varepsilon_T = \dfrac{\Delta W}{W}$ (Figure 12.8). The ratio of two strains, a dimensionless quantity, is called Poisson's ratio $v = \dfrac{\varepsilon_T}{\varepsilon}$, and depends on the material. It is usually expressed in microstrain, a deformation of 1/1,000,000. The strain gauge, used to measure the strain of objects in mechanical tests, as well as to detect force and other derived quantities such as acceleration, pressure, and vibration, consists of a grid of fine foil (Figure 12.9) or fine wire bonded to the carrier (Figure 12.10). The grid pattern maximizes the amount of foil that is subjected to strain. A sensor is attached to the device under test, and the resistance of the sensor varies linearly with the applied strain on the object. The strain gauge sensitivity or gauge factor is calculated as $\mathrm{GF} = \dfrac{\Delta R / R}{\Delta L / L}$ and is defined as fractional change of resistance to applied strain. The usual value of the gauge factor for foil type strain gauges is around 2.

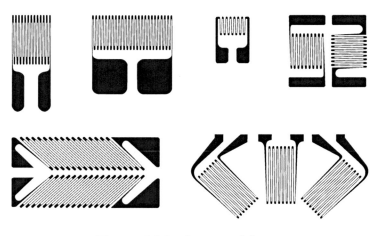

Figure 12.9 Strain gauge foil type

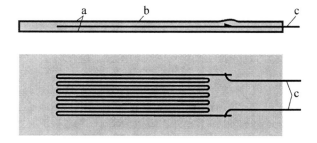

Figure 12.10 Strain gauge wire type

As strain gauges require measurement of small resistance changes, the Wheatstone bridge must be used. Strain gauges can occupy from 1 to 4 arms of the bridge to extend the sensitivity of the bridge by arranging the strain gauges in opposing directions. Figure 12.11 shows the half-bridge strain gauge configuration, with two strain gauges R_{G1} and R_{G2}, where strain gauge R_{G1} resistance increases with positive strain, while R_{G2} resistance decreases with the positive strain, thus doubling the sensitivity of the quarter-bridge configuration and canceling the temperature drift that is present in the quarter bridge configuration, in which a strain gauge occupies only one arm of the bridge.

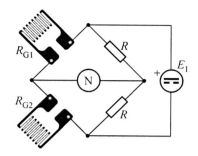

Figure 12.11 Half-bridge strain gauge bridge

12.2.5 Linear Voltage Differential Transformer (LVDT)

A linear voltage differential transformer is used to measure linear displacement (Figure 12.12).

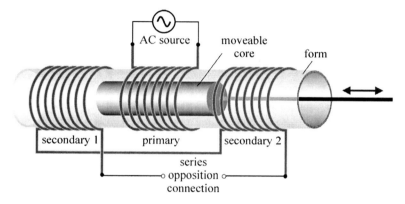

Figure 12.12 Linear voltage differential transformer

This instrument consists of three stationary coils (one primary and two secondary windings) and a movable magnetic core that can move back and forth without contact with the coils, ensuring a long period of use. The AC excitation voltage applied to the primary winding produces voltages in secondary windings through the core. When the core is positioned in the center, the voltages in secondary windings will cancel each other out, as they will be equal in amplitude, but 180 degrees out of phase. As the core moves from the center position, it will be more tightly coupled to one of the secondary windings, depending on the movement direction, thus producing an output voltage from which a displacement can be deducted (Figure 12.13).

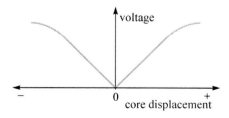

Figure 12.13 LVDT output voltage characteristics

12.2.6 Potentiometers as Displacement Sensors

The potentiometer as a position sensor consists of a brush that wipes along a resistive track of the potentiometer (Figure 12.14). The resistance from the brush to the beginning of the track is proportional with the position of the brush. To achieve the best possible linearity and low temperature drift, the resistance element is usually wound around the insulating core.

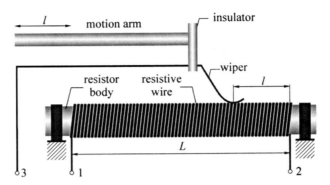

Figure 12.14 Potentiometer as position sensor

12.2.7 Accelerometers

An accelerometer is a sensor that measures acceleration, but can also be used to detect free fall and shock, movement, speed, and vibration. Accelerometers can be found with 1, 2, or 3-axis detection of acceleration. There are many different accelerometers defined by the principle of work: capacitive, piezoelectric, hall-effect, or heat transfer. The most common type of accelerometer is the capacitive accelerometer. The change of capacitance is caused by reducing or increasing the capacitor plate's gap. One plate is static and is placed on the accelerometer case, and the other is mobile under the influence of acceleration. An accelerometer can have several capacitors serially connected or in bridge. The principle of changing capacitance with acceleration is shown in Figure 12.15.

The most essential features of the accelerometer are: range, sensibility, and number of sensitive axis and output signal's independence of supply voltage. Range is defined with earth's gravity g (1g = 9.80665 m/s²); in today's accelerometers this can be from ±1.5g to more than ±100g. In some sensors, the range can be changed via a user application. Sensitivity is defined as the ratio of output voltage and the gravity of the sensor (V/g).

MEASUREMENT SYSTEMS

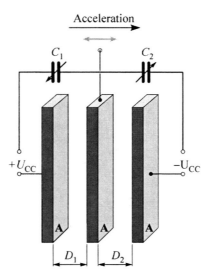

Figure 12.15 Change of capacitance in accelerometer

With high sensitivity sensors, it is possible to detect very small amounts of acceleration, vibration, and motion. To obtain the most resolution per degree of change, the accelerometer should be mounted with the sensitive axis parallel to the plane of movement where most sensitivity is desired. In Figure 12.16, the acceleration of a typical one-axis accelerometer in gs is shown as it tilts from -90° to +90°. The sensitivity for this axis is best for angles from -45° to +45°, and is reduced between -90° to -45° and between +45° to +90°. The sensitivity and precision of measurement can be enhanced if three axes are combined for a particular application.

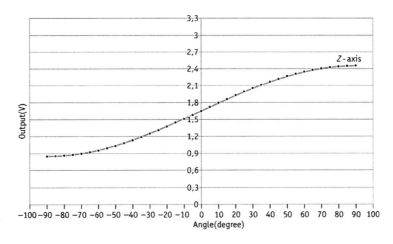

Figure 12.16 Nonlinearity of accelerometer's output.

12.2.8 Micro Machined Inertial Sensors (MEMS)

Micro machined inertial sensors (such as accelerometers) are used for many different applications, including navigation, impact detection, position, tilt, inclination, shock, vibration, motion detection, human tracking, etc. MEMS combine mechanical parts, electronics circuits, and even biochemical and chemical systems that give output signals dependent of sense activity, all integrated in one device.

MEMS technology is widely used, in areas ranging from robotics to agriculture. It is small in dimension, and uses low power at a low cost. MEMS sensors appear in fields such as optics, biomechanics, microfluidics, and radio-frequency. The advantages of MEMS are their small size, high sensitivity, low noise, and reduced cost, as these sensors are usually produced in batches.

12.3 TYPES OF SIGNALS

Signals received as output from the sensor can be categorized into two types:

- Digital
- Analogue

Useful information from the sensor can be contained in one or several parameters in the analogue signal:

- Level
- Shape
- Frequency

Useful information from the sensor can be contained in these parameters of a digital signal:

- Rate
- State

Digital signals can have only two states: High/ON and Low/OFF. A TTL (transistor to transistor logic) is usually used in digital signals, which means that 0-0.8 V is considered OFF, and 2-5 volts are considered ON. The area between +0.8 V and +2V is indeterminate and could be interpreted as either a logic high or a logic low, and should be avoided (Figure 12.17). The change of a digital signal with respect to time is measured by rate.

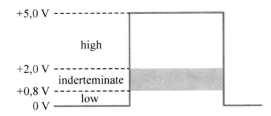

Figure 12.7 TTL signal

Analogue signals have three different parameters that can be measured:

- Level
- Shape
- Frequency

The analogue signal is continuous and can be in any state (voltage level) at any time; the voltage level is the most important parameter of the analogue signal to be measured. However, the shape of the analogue signal is also important as the shape factor can influence the RMS or peak value of voltage. The frequency of the measured analogue signal can be determined using various techniques such as zero crossing or Fast Fourier Transform (FFT).

12.4 SIGNAL CONDITIONING

The signals received from the sensor are not always suitable to pass on to the A/D converter directly, as the input of the A/D converter can only convert signals of certain voltage levels and quality. The signal from the sensor could be noisy, too small, or too large. Thermocouples, strain gauges, and microphones all produce a voltage in the millivolt range, which is too low for the A/D converter. Therefore, the sensor signal usually has to be conditioned (improved) before it can be converted to a digital form. In order to operate correctly, most sensors also need some sort of external hardware to perform their job, like the strain gauges that are placed as one resistance branch in the Wheatstone bridge. Sensors are classified as either active or passive. **Active** sensors, like resistance thermal devices (RTD), need an external source of **excitation** current. **Passive** sensors, such as thermocouples, that transform physical energy to electrical energy directly, generate output signals without the need for an excitation source.

Signal conditioning techniques can be divided into several major types:

- Excitation
- Amplification
- Linearization
- Isolation
- Filtering

12.4.1 Amplification

Amplification is used to increase the voltage level of a signal to make it fitting to be quantized by the A/D converter. The thermocouple is a good example, as its output is a voltage in the millivolt range. Such a low voltage is not suitable for most A/D converters, and needs to be amplified. Amplification is best performed by external amplification as close as possible to the sensor output source, and not in the A/D card, as the A/D card will also amplify the noise picked up by lead wires that connect the sensor to the A/D converter (Figure 12.18).

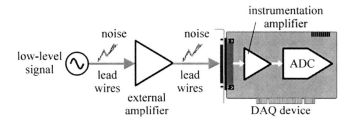

Figure 12.18 Amplification as close as possible to the signal source to avoid picking up the noise by the lead wires

12.4.2 Excitation

Active sensors are excited using either a controlled current or voltage. Strain and pressure sensors are usually excited with constant-voltage excitation, while constant current excitation is used for resistive sensors such as RTDs or thermistors. Current excitation is preferable to voltage excitation in noisy environments due to its better noise immunity.

DC excitation is simpler to implement and usually low cost, but it is susceptible to noise and offset errors. AC excitation, even though it is more difficult to implement, offers better immunity to noise, offsets, and effects of the parasitic thermocouples.

12.4.3 Linearization

Linearization is needed when the output of the sensor is non-linear, which is the case for almost all temperature sensors except IC temperature sensors (Figure 12.19). In the past, complex analogue conditioning circuits were designed to correct for the sensor nonlinearity, as non-linearity causes the largest measurement errors in temperature measurement. Today, sensor outputs are digitized by high speed and accurate DAQ cards, so linearization can be obtained by using look-up tables from the user application or tables that are stored in the microcontroller memory, thus simplifying the requirements on the analogue circuitry.

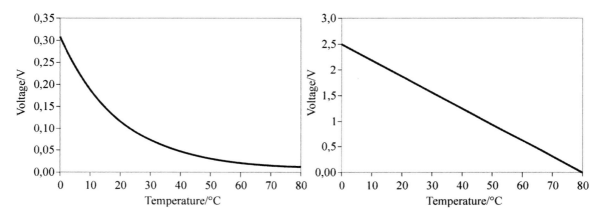

Figure 12.19 Sensor non-linear output is shown on the left (typical for temperature sensors). The output after the output linearization is shown on the right.

12.4.4 Isolation

Isolation is used to pass the signal from the source to the measurement system without physical connection.

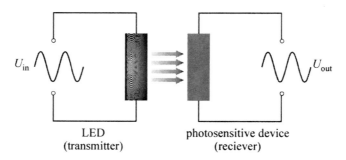

Figure 12.20 Optocoupler isolation

In this way, problems caused by improper grounding do not affect the measurement. In addition, isolation blocks high voltage surges and rejects high voltage transients, thus protecting the expensive measuring equipment and operator. For the purpose of isolation, several techniques are used, such as transformers and optical (Figure 12.20) or capacitive coupling.

12.4.5 Filtering

Filters reject unwanted noise within a certain frequency range. For example, DAQ cards are usually susceptible to 50 Hz or 60 Hz noise from power lines and other environmental sources. With filtering, this noise is eliminated. Also, anti-aliasing filters eliminate the problem that arises when the signal is undersampled. A filter with a cutoff frequency of less than half of the sampling rate is used for that purpose. Filtering can be also performed digitally by a user application or data analysis software, such as LabWIEW™, which offers a wide selection of filters.

12.4.6 Comparison of Signal Conditioning Requirements for Different Sensors

In Table 12.1, signal conditioning requirements for some of the most popular sensors are shown. Some sensors require extensive signal conditioning, while others do not require signal conditioning at all.

Table 12.1 Electrical Characteristics and Basic Signal Conditioning Requirements of Common Transducers

Sensor	Electrical Characteristics	Signal Conditioning Requirement
Thermocouple	Low-voltage output Low sensitivity Nonlinear output	Reference temperature sensor (for cold-junction compensation) High amplification Linearization
RTD	Low resistance (100 ohms typical) Low sensitivity Nonlinear output	Current excitation Four-wire/three-wire configuration Linearization
Strain gauge	Low resistance device Low sensitivity Nonlinear output	Voltage or current excitation High amplification Bridge completion Linearization Shunt calibration
Current output device	Current loop output (4 -- 20 mA typical)	Precision resistor
Thermistor	Resistive device High resistance and sensitivity Very nonlinear output	Current excitation or voltage excitation with reference resistor Linearization
Active Accelerometers	High-level voltage or current output Linear output	Power source Moderate amplification
AC Linear Variable Differential Transformer (LVDT)	AC voltage output	AC excitation Demodulation Linearization

12.5 DATA ACQUISITION HARDWARE

After a physical phenomenon has been converted to a measurable signal, it is acquired by the data acquisition system (DAQ). The data acquisition system is made of a cable, DAQ card with A/D converters, and a computer. DAQ cards that are inserted into the PC expansion slot are usually delivered with an external terminal block containing screw terminals for connecting the signals and a cable for connecting the terminal block with the DAQ device in the PC (Figure 12.21). In this way, the signal is transported to the DAQ card in the PC via a shielded cable, and not directly from the sensor; it is thus better protected from the noise. Also, terminal blocks are easier to reach than DAQ cards inside the PC.

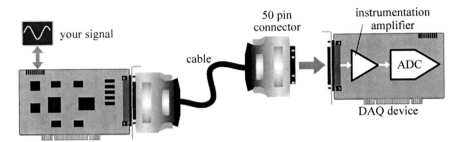

Figure 12.21 PC based DAQ card with external terminal block

Some terminal blocks even have extra features, such as cold-junction compensation necessary to connect a thermocouple. DAQ devices are available for several popular buses used in the PC industry today (USB, PCI, PXI, etc.) Today, variants with Ethernet and wireless possibilities are also available.

DAQ devices can have many different inputs and outputs. The most important element of DAQ is the analogue input, but the analogue output, digital I/O, and counters/timers can also be part of the DAQ device.

12.5.1 DAQ Characteristics

There are several important characteristics of analogue inputs that the user should be acquainted with before the device can be used to achieve the best possible measurement results. These are:

- Resolution
- Range
- Gain
- Code Width
- Grounding issues.

Resolution is the number of bits which a DAQ device uses to represent the signal. As each binary number represents a certain voltage level, the resolution may be expressed in volts to see how many voltage levels can be measured. For example, a 16 bit resolution has $2^{resolution} = 2^{16} = 65,536$ levels, and for the voltage range of 10 V, the resolution in volts is approximately **10 V/65536=152 µV**. The higher the resolution, the smaller voltage increments can be measured (Figure 12.22), and the digital representation of analogue input signal will be more precise. Today, 16-bit, 24-bit, and even 32-bit ADC converters are available on the market for different purposes.

MEASUREMENT SYSTEMS

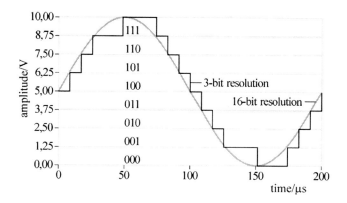

Figure 12.22 ADC with 3-bit resolution and 16-bit resolution (5 kHz wave)

The analogue output specifics are the choice of internal or external voltage reference, and the choice between the bipolar or unipolar signal.

Range refers to the minimum and maximum voltages that an ADC can digitize.

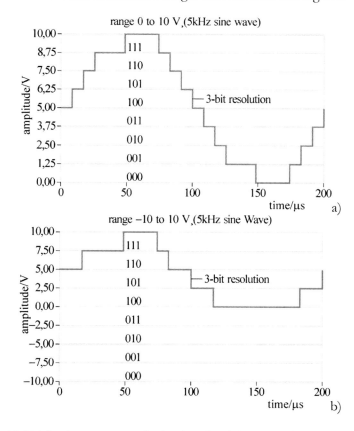

Figure 12.23 The importance of selecting the right range for the input signal

207

Most DAQ devices have a range from 0 to 10 V, or from -10 V to 10 V; however, for different purposes the ADC range can vary greatly. Some of the DAQ cards have selectable ranges.

If possible, the user should select the range to closely match the range of the input analogue signal. In this way, the available resolution will yield the best representation of the input signal (Figure 12.23a). Figure 12.5a shows the correct range (0 to 10 V) to maximize the available resolution, as it uses all available eight levels. In Figure 12.23.b, only four levels are used for the same input signal, as the range is from -10 V to 10 V.

Gain is a measure of the amplifier for increasing the amplitude or power of the input signal, and is defined as the ratio of the output to the input signal. This topic was discussed in Chapter 7. Gain is achieved with the instrumentation amplifier that is part of the DAQ card. The user selects the minimum and maximum input values and range, and the maximum gain is then indirectly set by the measurement system to achieve the best possible representation of a signal and use of the chosen range. The gain settings are 0.5, 2, 5, 10, 20, 50, and 100 for most devices. They can be seen in Figure 12.24.

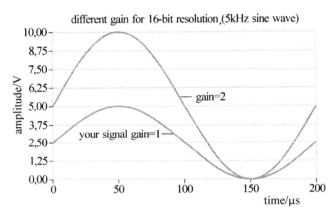

Figure 12.24 Gain settings for input signal with 0-5 V limits, range settings 0-10 V and gain settings set to 2 to maximize the use of the available resolution

Code width is the smallest change in the signal a DAQ card can detect. It is determined by resolution, range, and gain:

$$\text{Code width} = \frac{\text{range}}{\text{gain} \cdot 2^{\text{resolution}}} \qquad [12.3]$$

Smaller code width means better representation of a signal. To enhance the representation of a signal, it is therefore necessary to increase the range, gain, or resolution. For example, a 16 bit DAQ card with gain=2 and 0-10 V range has a code width equal to 76.3 µV.

12.5.2 Grounding Issues of DAQ Measurement System

Proper **grounding** of the measurement system is necessary for good measurement results. First, it is necessary to know the input signal grounding configuration. The input signal can come from a **grounded** source (power supplies, signal generators, etc.) or from **floating** sources (batteries, thermocouples, transformers, etc.), as seen in Figure 12.25.

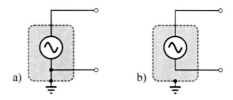

Figure 12.25 Signal source can be grounded (a) or floating (b)

The DAQ card can have three possible grounding configurations (Figure 12.26):

- Differential
- Referenced single ended (RSE)
- Non-referenced single ended (NRSE)

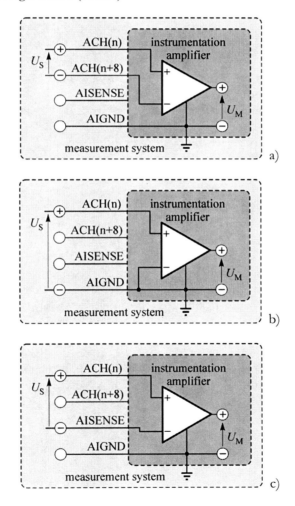

Figure 12.26 Grounding configurations for the measurement system

In differential configuration, neither positive nor negative input of the DAQ instrumentation amplifier is connected to the system ground (Figure 12.26a). In the referenced single-ended configuration, the measurement system connects negative input of the instrumentation amplifier to the system ground (Figure 12.26b). In the non-referenced single-ended configuration that is offered by some manufacturers, the measurements are still made with respect to a common reference as in the referenced single-ended configuration, which can vary with respect to the system ground (Figure 12.26c).

The best configuration for both grounded and floating signal sources is the differential, as it will reject common mode voltage (voltage common to both inputs) which produces measurement errors. The only drawback is that the number of available channels will be cut in half. The common-mode voltage V_{CM} is defined as $V_{CM} = (V_+ + V_-)/2$, where V_+ is the voltage at the noninverting terminal of the measurement system with respect to the measurement system ground, and V_- is the voltage at the inverting terminal of the measurement system with respect to the measurement system ground. The common-mode rejection ratio (CMRR) is the factor that determines the ability of the circuit to reject the common-mode voltage (Figure 12.9). The CMRR in dB is defined as follows:

$$\text{CMRR (dB)} = 20 \log (\text{Differential Gain}/\text{Common-Mode Gain}). \qquad [12.4]$$

The RSE is not recommended for grounded sources, as the ground potential at signal source and measurement system may vary. In that case, the created ground loop will add AC and DC noise in the measurement system, causing measurement errors. In floating signal sources, the RSE configuration can be used to maintain the channel count, but the common-mode voltage will not be rejected. The NRSE configuration can be used for grounded and floating signal sources when all available channels must be used; however, it will not reject the common mode voltage. Also, for floating signal sources, a bias resistor must be utilized to provide the path to ground for any bias current in the instrumentation amplifier.

12.5.3 Sources of Noise in the DAQ Measurement System

Measuring analogue signals with a data acquisition device is not always simple, and the operator must be aware of different noises and disturbances that can affect measurement accuracy. It is important to know the nature of the signal source, configure the DAQ system, and use the appropriate cabling to ensure noise-free and accurate measurements. Noise sources can be divided into internal and external. Internal noise sources arise from the temperature change, such as Johnson noise. External noise in DAQ measurement systems can arise from different causes, all the way from signal source to measurement system (Figure 12.27). Typical sources of noise are power lines (50/60 Hz noise), computers, monitors, pulsing digital lines, and high-voltage switching. Noise reduction techniques focus on minimizing all of these factors, as noise is amplified together with the measured signal. Noise is transferred to the signal by coupling channels that can be divided into conductive, capacitive, inductive, and radiative coupling. Common-mode interference can also cause measurement errors.

MEASUREMENT SYSTEMS

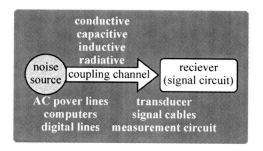

Figure 12.27 Noise source in DAQ measurement system

Conductive coupling is created because conductors have finite impedance. It can be minimized if ground loops are eliminated. Also, separate grounding must be provided for all signal sources, especially for low level signal sources and high-level sources passing large currents. In series grounding (Figure 12.28a), large currents from one signal source (I_3) can create considerable voltage drops on the ground impedances of other low level signal sources (R_1 and R_2). A correct grounding is shown in Figure 12.28b.

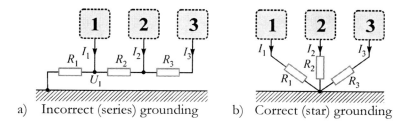

a) Incorrect (series) grounding b) Correct (star) grounding

Figure 12.28 Conductive coupling and grounding

Capacitive coupling in the measurement system results from electric fields from nearby circuits (Figure 12.29).

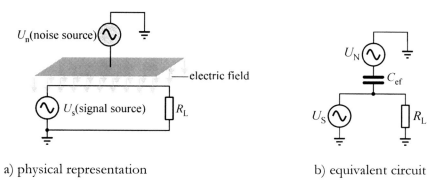

a) physical representation b) equivalent circuit

Figure 12.29 Capacitive circuit

In equivalent circuits, the capacitance C_{ef} represents an electric field coupling between two circuits. As the capacitive coupling is proportional to the area of overlap and inversely proportional to the distance between the two circuits, the capacitive coupling can be minimized if the overlap is kept at min-

211

imum, and if the circuits are separated. The capacitive coupling can be also minimized by shielding, which offers another path for the induced current noise (equivalent capacitance multiplied by the changing electric field) (Figure 12.30). The shield must be placed between the conductors that are capacitively coupled and connected to ground only at the signal source. If the shield is grounded on the measurement system, as well as on the signal source, ground loop current occurs.

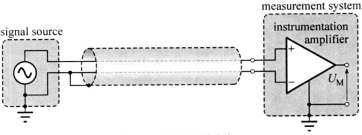

Figure 12.30 Shielding

Inductive coupling results from time-varying magnetic fields generated by currents in nearby noise circuits. These magnetic fields, through mutual inductance with signal circuits, induce noise voltage (Figure 12.31).

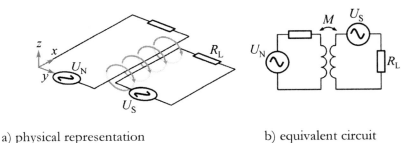

a) physical representation   b) equivalent circuit

Figure 12.31 Inductive coupling

Inductive coupling can be minimized by separating two circuits or minimizing the signal loop area, which can be done by using twisted pair conductors (Figure 12.32).

Figure 12.32 Twisted pair conductors to minimize inductive coupling

Shielding can also be used, but it is not as effective as capacitive coupling. The effectiveness of shielding depends on the material used; it is mostly ineffective at low frequencies (such as the power line frequency).

Radiative coupling results from sources such as radio and TV broadcast stations. This high frequency noise is coupled to low frequency measurement systems through the rectification process in nonlinear junctions of integrated circuits. Radiative coupling is reduced by using passive R-C low-pass filters at the receiver end.

Common-mode noise is often referred to as common-mode voltage (CMV), which is present at both inputs of an analogue circuit with respect to the analogue ground. The major source of common-mode noise is the difference in potential between two physically remote grounds. Figure 12.34 shows the measurement of voltage signal U_i, with digital voltmeter (DVM) analogue to the digital converter (ADC), where the LO input is connected to the internal shield which is grounded. The conductors have resistances R_a and R_b, and the input impedance of ADC or DVM is presented as Z_{ul}. The common-mode noise is presented as U_{SZ}, causing the ground loop current I_{SZ}. That is why grounding the LO input is not recommended on both ends of the measurement system.

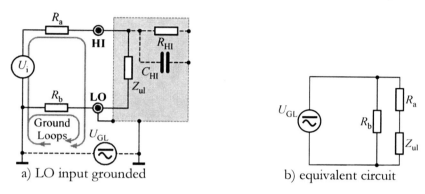

Figure 12.34 Ground loop current and common-mode noise

Common-mode noise rejection is better if floating input is used (Figure 12.35), or if input is separated from the grounded inner shield of the circuit by high-ohm impedance Z_{LO}. In this way the ground loop current will be greatly reduced.

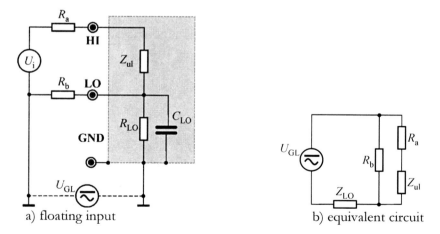

Figure 12.35 Floating input provides better common-mode rejection ratio

The best way to reject common-mode noise is to use the GUARD shield, an additional shield between the low and ground terminals. It increases LO terminal to ground resistance and decreases capacitance. The GUARD shield increases the rejection ratio of common-mode noise; however, if properly connected as in Figure 12.36, it will provide the path for common-mode noise current away from the HI and LO connector resistances R_a and R_b. In this way, the CMRR can be as high as 160 dB for DC noise, and 120 dB for AC noise. The guard connection is available with high-quality digital voltmeters or with ADC of high resolution (24 bit or higher).

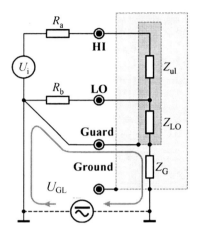

Figure 12.10 GUARD connector

12.6 EMBEDDED AND SYSTEM ON CHIP (SoC) MEASUREMENT SYSTEM

Embedded systems are computers designed to perform several functions, and they usually have small dimensions. In recent years, there have been many new measurement systems realized with the help of microcontrollers that practically have the whole measurement system realized on a single chip. The functionalities include analogue to digital converters, program and data storage, signal conditioning (rectifiers, filters, digital signal processing), excitation (voltage and clock references, current and frequency sources), analogue and digital outputs, timers and counters, as well as transceivers such as ZigBee or Bluetooth. Some commercial solutions provide plug and play functionalities that include an interface, such as USB with an application programming interface (API). Manufacturers can also include software support that simplifies the installation and use of such sensors in many applications and programming environments, such as LabVIEW™.

12.6.1 Smart Sensors

A smart sensor is capable of not only converting a physical quantity into an electrical signal (voltage or current), but also of converting a measured value into a digital format in the units of the measured quantity, and then transmitting that measured information to a computer or other controller, wireless or wired. The Instrumentation and Measurement Society's Sensor Technology Technical Committee has been developing an IEEE 1451 set of smart transducer interface standards for connecting transducers (sensors or actuators) to microprocessors, instrumentation systems, and control/field networks. The most important achievement of this standard is the definition of TEDS (Transducer Electronic Data Sheet), which every transducer should possess to enable its identification, calibration, correction data, and other relevant data, such as manufacturer and serial number.

12.6.2 Wireless Sensor Networks

A sensor equipped with wireless communication, a microcontroller, and an energy source can be employed in wireless sensor networks (WSN) (Figure 12.37). Wireless sensor networks are spatially distributed sensors used to gather environmental or physical information for post-processing and analysis. The scientific field of wireless sensor networks has been extensively researched in recent years.

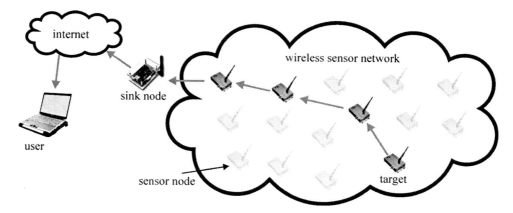

Figure 12.37 Wireless sensor networks

Selected bibliography:

1. Nate Mehring, *Introduction to Engineering Technology*, Texas A&M University, http://etidweb.tamu.edu/ftp/ENTC250/Nate/LabVIEW6.ppt, Retrieved October 15th, 2010
2. Britannica Online Encyclopedia, *Measurement System*, http://www.britannica.com/EBchecked/topic/1286365/measurement-system, Retrieved October 15th, 2010
3. Syed Jaffar Shah, *Field Wiring and Noise Considerations for Analogue Signals*, National Instruments Application Note 025, http://cires.colorado.edu/jimenez-group/QAMSResources/Docs/Noise-analog.pdf, Retrieved October 9th, 2010
4. Albert O'Grady, "*Transducer/Sensor Excitation and Measurement Techniques*", Analogue Dialogue, Volume 34, Number 5, August-September, 2000, http://www.analog.com/library/analogDialogue/archives/34-05/sensor/index.html, Retrieved October 15th, 2010
5. *Signal Conditioning Fundamentals for Computer-Based Data Acquisition Systems*, NI Developer Zone, http://zone.ni.com/devzone/cda/tut/p/id/4084, Retrieved October 15th, 2010
6. Walt Kester, James Bryant, Walt Jung, *Temperature Sensors*, http://www.analog.com/static/imported-files/tutorials/temperature_sensors_chapter7.pdf, Retrieved October 15th, 2010
7. Gerald Jacob, *A Primer on Signal-Conditioning Issues*, Evaluation Engineering, http://www.evaluationengineering.com/index.php/solutions/instrumentation/a-primer-on-signal-conditioning-issues.html, Retrieved October 15th, 2010
8. Luka Ferković, *Measurement Methods (in Croatian), Lecture notes*, http://www.fer.hr/_download/repository/MM_-_Tema9.pdf ,
9. *The RTD*, http://www.omega.com/temperature/z/thertd.html, Retrieved October 15th, 2010
10. *Measuring Strain with Strain Gauges*, NI Developer Zone, , Retrieved October 15th, 2010
11. Mark Whittington , *Measuring Strain, Position, and Acceleration*, NI Week, 1999

12. Scott Ziffra and Michel Haddad, Improving *Signal Quality via Signal Conditioning Techniques*, NI Week 1998
13. Luka Ferković, *Measurement Methods (in Croatian), Lecture notes*, http://www.fer.hr/_download/repository/MM_-_Tema10.pdf, Retrieved October 15th, 2010
14. *USB Sensors and Peripherals*, http://www.toradex.com/En/Products/USB_Sensors_and_Peripherals, Retrieved October 9th, 2010
15. *MAXQ7667, 16-Bit, RISC, Microcontroller-Based, Ultrasonic Distance-Measuring System*, http://www.maxim-ic.com/datasheet/index.mvp/id/6198, Retrieved October 15th, 2010
16. Jerry Gaboian, *A Survey of Common-Mode Noise*, Application Report, SLLA057, December 1999, http://focus.ti.com/lit/an/slla057/slla057.pdf, Retrieved October 15th, 2010
17. *TinyTIM™ (Tiny Transducer Interface Module), Bluetooth Smart Sensor Module*, Smart Sensor Systems, Inc., www.SmartSensorSystems.com, Retrieved October 15th, 2010
18. F. L. LEWIS, *Wireless Sensor Networks, Smart Environments: Technologies, Protocols, and Applications*, http://citeseerx.ist.psu.edu/viewdoc/download?doi=10.1.1.98.794&rep=rep1&type=pdf, Retrieved October 15th, 2010
19. Malarić, Roman; Hegeduš, Hrvoje; Mostarac, Petar, *Use of Triaxial Accelerometers for Posture and Movement Analysis of Patients*, Advances in Biomedical Sensing, Measurements, Instrumentation and Systems, Springer Verlag, 2009. pp. 127-143
20. Pavel Lajšner, Radomir Kozub, *Wireless Sensing Triple-Axis Reference Design* (ZSTAR), Freescale Application Note 3152
21. *Insulator Seal, Thermocouple Temperature-vs-Voltage*, http://www.insulatorseal.com/searchs/doc/Thermo-Graph.htm, Retrieved October 15th, 2010

ABOUT THE AUTHOR

Roman Malarić was born on February, 1971 in Zagreb, Croatia. He received a B. S. from the University of Zagreb, Croatia in 1994 in the field of radio communications. He earned M. Sc. and Ph.D. degrees from the same university, in 1996 and 2001 respectively. In 2004, he become an assistant professor at Zagreb University, Faculty of Electrical Engineering and Computing, Department of EE Basics and Measurement, where his primary interests are the development of precise measurement methods for electrical quantities, and the automatization of measurement methods by computers and virtual instrumentation. In 2008, he became an associate professor in the same faculty. He is involved with the Primary Electromagnetic Laboratory of Croatia, and contributed to the accreditation of the laboratory by the DKD in 2005. He has written more than 50 scientific articles for conferences, and published 15 articles in scientific journals. He has been a member of organizing committees of several IMEKO scientific conferences and symposiums. He is a member of IEEE and its technical committees, and has served as a reviewer for IEEE Transactions of I&M for several years.

INDEX

1-phase, 50
3-phase, 50

A/D Converter
    Dual slope, 3, 121
    Parallel, 3, 123
    Successive approximation, 3, 122
    Voltage to time, 3, 120
accelerometers, 4, 195, 200, 211
accuracy, 3, 5, 11, 14, 23, 31-33, 45, 47, 49, 50, 58, 63, 65, 69, 70, 78, 82, 83, 91, 92, 95, 97, 115, 117, 121, 122, 125, 126, 130, 137, 140, 146, 191, 205
accuracy class, 47, 49, 50, 82, 126
active guard, 95
active power, 138
aliasing error, 114
ammeter, 49, 68, 132, 136-141, 145, 147, 149
amplification, 4, 198
amplifier
    frequency bandwidth, 88
    negative feedback, 88-91, 99
    symbol, 88
analogue signals, 205
analogue to digital converter
    PWMDAC, 71, 72, 220
analog-digital converters, 118
analogue-digital converters, 3, 113, 118, 120-124, 126, 198, 201, 220
anode
    focusing, 105, 111
apparent power, 138
arithmetic mean, 13, 15, 16
Aron connection, 143, 144
asynchronous switching mode, 108, 111
atomic clock, 6

block diagram, 71, 104, 114, 121, 178-180
bridge
    AC, 67, 68
    balancing, 68
    Glynn, 159
    GPIB-USB, 170
    Maxwell, 154
    Schering, 156, 157, 158
    sensitivity, 2, 65
    Thompson, 2, 66
    Wagner, 158
    Wheatstone, 2, 66
    Wien, 156
    with Variable Inductance, 153
    with Variable Inductance, 4
    with Variable Inductance, 153

calibrator, 71, 72
    AC, 70, 72
    DC, 4, 70
capacitive coupling, 206
capacitors
    decades, 37
    equivalent Circuit, 1, 35
    loss angle, 36
    plate capacitors, 36
    standards, 1, 36
    Thompson-Lampard standard, 5, 36
cathode, 4, 104-106, 108, 114
cathode ray tube, 4, 104, 105, 106, 114
chopper, 97, 98
Chubb method, 135
coaxial cable, 106, 107
code width, 203
common-mode noise, 208
comparator, 123, 124
compensation, 4, 41, 43, 63, 68, 69, 83, 85, 122, 131, 133, 146, 189, 200, 201
conductive coupling, 206
confidence limits, 17, 18, 19
contact resistances, 66, 145
correction, 12, 31, 137, 209
coulonmeters, 93
crest factor, 55, 125
current clamps, 132
cut-in voltage, 53

daisy chain, 168, 169
damping torque, 47
DAQ, 5, 174, 189, 190, 199-201, 203-206, 220
degree of attenuation, 47
delay line, 106, 108

digital multimeter, 4, 117-119, 182, 184, 186, 220
Digital multimeter, 184
digital sampling AC converter, 125
digital signals, 197
Digital signals, 197
digital storage oscilloscope, 4, 220
digits, 117, 126
displacement Sensors, 4, 195
Displacement Sensors, 195
distribution
    normal, 15, 16
    rectangular, 19
divider
    inductive, 133
    resistive, 132, 133
drift, 2, 6, 30, 32, 95, 97, 98, 194, 195

effective value, 54-58, 72, 99, 101-103, 124, 126, 198, 221
electricity meter
    electronic meter, 128
    induction meter, 58, 59, 60, 128
electrometer, 130, 131
electron beam, 106, 111
electronic voltmeters
    AC, 98
    DC, 97
ethernet, 4, 171, 173, 182, 201
excitation, 4, 198, 210

factor of stabilization, 42
fall of potential, 150
filtering, 35, 200
FireWire, 4, 167, 172, 173
front panel, 105, 111, 176, 178-180
function generator, 39, 184

gain, 87-92, 94, 95, 96, 203
general mean, 15
General Purpose Interface Bus, 4, 167, 168, 220
graphical programming, 4, 177
Greatz circuit, 100
ground loop, 205, 206, 207, 208
grounding, 112, 139, 150, 200, 203, 204, 206, 208
guard, 3, 95

Hall effect, 25, 34
Hamon resistor, 1, 33, 44, 45
Hewlett Packard Interface Bus, 167, 220
horizontal deflection system, 104, 106, 108, 109, 112
HS488, 4, 170, 173, 186

IICP488, 173
indirect Measurements, 1, 20
inductive coupling, 207
inductors
    Equivalent Circuit, 1, 38
    quality factor, 38, 152
    standards, 38, 39
influential factors, 50
input capacitance, 119
input impedance, 91
input offset current, 119
input resistance, 91, 96, 97, 101, 119, 140
instrument control, 166
instrument drivers, 4, 176, 184, 185, 186
insulation resistance, 3, 148, 163
interface buses, 4, 173
interfaces, 166, 167, 168
isolation, 33, 50, 199, 200

Kelvin-Varley divider, 70, 133

LabVIEW™, 176, 177, 180, 182-186, 220
LAN Extensions for Instrumentation, 4, 171, 220
LCR meter, 160
limits of error, 3, 19, 21, 22, 48, 49, 65, 126, 140, 174
Linear Voltage Differential Transformer, 4, 194, 220
linearization, 199
loss angle, 36, 67, 156, 160, 161, 162

mean value, 13, 17, 18, 29, 31, 32, 33, 51, 54, 56, 57, 99, 100, 101, 124, 161, 162
measurement errors, 3, 5, 11, 12, 95, 149, 191, 199, 205
    absolute errors, 11, 12
    correction, 12, 31, 137, 209
    grave errors, 12
    random errors, 13, 15, 16, 18, 20
    relative errors, 11, 12

systematic errors, 12, 13, 16, 18, 19
measurement method
  capacitance, 155, 156, 157, 158, 159
  current, 130
  current measurement, 6, 12, 52, 75, 132
  DC power, 3, 136, 137
  Earth Resistance, 3, 150, 163
  frequency, 125, 162, 163
  high voltage, 132, 135
  High-Ohm resistance, 3, 148
  inductance, 152, 153
  insulation resistance, 3, 148, 163
  large current, 131, 132
  power, 135, 139, 140, 141, 143, 144
  resistance, 3, 144, 145, 146, 148, 149, 152, 153, 155
  small current, 130
  soil resistivity, 3, 151
  time, 162
measurement range, 49, 52, 75, 96, 126, 138
measurement repeatability, 14
measurement reproducibility, 14
measurement systems, 188
measurement uncertainty, 18
measuring instrument, 3, 4, 11, 14, 18, 46-50, 52, 55, 70, 75, 81, 87, 97, 117, 118, 146-149, 166, 171, 177
measuring instruments
  analogue, 4, 46, 48, 49, 50, 52, 56, 57, 63, 97, 101, 117
  digital, 117, 118
  electrodynamic, 57, 101, 138
  IMCPM, 2, 51, 52, 53, 55, 56, 57, 220
  moving iron, 50, 56, 57, 101
  pointing device, 46, 47, 48, 49, 51, 117
  scale, 12, 47, 48, 49, 50, 51, 56, 57, 58, 102, 117
  uncertainty, 2, 48
micro machined inertial sensors, 4, 196, 220
modular instruments, 171, 174, 175, 177, 178

noise, 23, 95, 121, 126, 197-201, 205-208
null-instruments, 131

Ohm-Meter, 3, 146, 147
oil bath, 26, 27
operational amplifiers, 4, 91, 94-97, 99, 101, 102
  difference amplifier, 95

differentiator amplifier, 93, 94
instrumentation amplifier, 95, 203, 205
integrating amplifier, 16, 93, 102, 121, 160, 161, 162
inverting, 91, 92, 93, 96, 99, 119, 123, 160, 205
logarithmic amplifier, 93
non-inverting, 91, 92, 93, 96, 119, 123
summing amplifier, 21, 92
oscilloscope, 63, 68, 83, 104-111, 113-116, 174, 220
oscilloscope probe compensation, 107

PCI eXtensions for Instrumentation, 4, 175, 220
peltier coolers, 27
Peripheral Component Interconnect, 4, 174, 175, 220
phase error, 77, 82, 83, 84, 85
power supply, 96, 110, 117, 184
precision, 6, 13, 14, 25, 31, 64, 90, 91, 97, 99, 119, 122, 124, 196
prefixes, 8, 9, 10
probe, 106, 107, 112, 131, 132, 150, 151
programmable instruments, 4, 181, 221

Quantum Hall effect, 1, 6, 34, 40, 44

radiative coupling, 207
range, 32, 35, 41, 44, 47, 49, 50, 52, 53, 75, 88, 96, 97, 108, 117, 119, 126, 130, 138, 147, 168, 172, 182, 188-190, 195, 198, 200-203
rectifier, 39, 53, 54, 55, 56, 98, 99, 135, 209
reference values, 50
Resistance Thermal Devices, 4, 190, 221
resistors
  bifilar winding, 24
  decades, 117, 154
  equivalent circuit, 23
  layer resistors, 24
  potentiometer, 24, 43, 44, 68, 69, 70, 95, 112, 195
  Quantum Hall resistance standard, 1, 6, 34, 40, 44
  resistance standards, 11, 19, 25, 26, 27, 28, 29, 30, 32, 35, 66, 146
  slide resistor, 24, 43
  variable resistor, 24, 83, 153, 154, 155

resolution, 14, 117, 118, 126, 174, 182, 196, 201-203, 209
ring comparison, 31, 32
RS-232, 4, 167, 170, 173, 178, 186, 221
RS-485, 4, 170, 171, 178

safe limits of error, 21
sampling frequency, 114
sawtooth voltage, 55, 104, 106, 108-113, 120
sensitivity, 14, 49, 53, 54, 56, 63-65, 97, 121, 172, 193, 196, 197, 200
sensor, 96, 176, 188, 189, 193, 195, 197, 198-201, 209, 210, 221
shape factor, 55, 56, 57, 198
shielding, 149, 157, 206, 207
signal conditioning, 4, 126, 172, 188, 200, 209
spectrum analyzer, 184
Spectrum analyzer, 184
spherical gaps, 133, 134
Spherical gaps, 133, 134
standalone Instruments, 4, 173
Standalone Instruments, 173
standard deviation, 13-15, 17, 19, 20-22, 31
statistical error limits, 22
strain gauges, 4, 192, 210
synchronous switching mode, 108, 111

thermistors, 4, 192
thermocouple, 72, 102, 189, 190, 191, 198, 201
thermoelectric voltage, 145
three voltmeter method, 139
three-phase symmetric system, 141, 143
time division multiplier, 127, 128
torque, 46-48, 51, 56-59
traceability, 6, 29, 36
transconductance amplifier, 87
transducer, 96, 104, 126, 128, 188, 209
transformer
  accuracy class, 78, 82
  accuracy testing, 2, 83, 84, 85
  capacitive, 79
  current, 75, 76, 80-83, 84, 86, 132, 159
  errors, 4, 77
  ideal, 75, 76, 77
  variable, 86
  voltage, 75, 76, 77, 78, 79, 133
transimpedance amplifier, 87

trigger, 109, 110, 112, 180
true RMS, 101
true value, 11, 12, 13, 14, 15, 16, 19
TTL signal, 197
twisted pair, 207

units
  base units, 3, 4
  binary, 8, 9, 10, 124, 168, 174, 176, 180, 181, 201-203, 209
  BIPM, 2, 220
  calculabe standards, 5, 36
  calculable standards, 5, 36
  capacitance, 5
  CGPM, 2, 220
  CIPM, 2, 220
  definitions, 2
  derived, 3, 4
  importance, 2, 10
  non-SI units, 8
  realization, 5
  resistance, 5
  SI, 1, 2
  time, 2
  unit of force, 2, 3
  voltage, 3
Universal Serial Bus, 4, 172, 221
universal unstrument, 2, 56
USBTMC, 172, 221
USTMC Class, 4, 172

vertical deflection system, 104, 106, 108, 110
virtual instruments, 176, 177, 178
VISA, 4, 182, 183, 184, 221
VMEbus eXtensions for Instrumentation, 4, 174, 221
voltage
  standards, 1, 5, 6, 39, 40, 41, 43, 220
voltage follower, 94, 95
voltage source
  batteries, 39
  DC, 39
voltage sources
  electronic, 39
  signal generators, 39
voltage to current converter, 96
voltmeter, 12, 25, 26, 49, 50, 52, 58, 69, 75, 79, 97-103, 117, 126, 130, 136-140, 142, 144-149, 151-153, 155, 191, 208, 220

voltmeter-ammeter method, 145, 148, 149,
   152, 153, 155
VXI, 4, 167, 171, 174-176, 178, 182, 186, 221
VXI-11, 4, 171, 186

wattmeter, 58, 127, 138, 139, 142, 144, 153
   connecting, 139
waveform, 39, 54, 55, 101
Wehnelt cylinder, 105, 111
wireless sensor networks, 5, 209, 211

Lightning Source UK Ltd.
Milton Keynes UK
UKOW020203130712

195860UK00003B/27/P